PLANTA SAPIENS

PLANTA SAPIENS

THE NEW SCIENCE OF PLANT INTELLIGENCE

Paco Calvo
with Natalie Lawrence

W. W. NORTON & COMPANY
Celebrating a Century of Independent Publishing

For information about special discounts for bulk purchases,
please contact W. W. Norton Special Sales at
specialsales@wwnorton.com or 800-233-4830

Manufacturing by Lakeside Book Company
Production manager: Anna Oler

ISBN 978-0-393-88108-0

W. W. Norton & Company, Inc.
500 Fifth Avenue, New York, N.Y. 10110
www.wwnorton.com

W. W. Norton & Company Ltd.
15 Carlisle Street, London W1D 3BS

1 2 3 4 5 6 7 8 9 0

For Anabel

Verde que te quiero verde

Federico García Lorca

CONTENTS

But one never knows where to find them. The wind blows them away.
They have no roots, and that makes their life very difficult.

Antoine de Saint-Exupéry, *The Little Prince*

PREFACE

I have spent many years seeking to understand the experiences of organisms that are very different from us: to uncover the nature of plant intelligence. It's no small feat. The scientific work is far from done, but what we have found out so far just about shows us how much more there is to discover. This book is the culmination of two decades of passionate exploration into a rich and alternate world that exists alongside our own.

My venture began in 2006 when I read a book about the neuronal aspects of plant life edited by three scientists, František Baluška, Stefano Mancuso and Dieter Volkmann. This might sound strange: plants do not have neurones, after all. I had never considered plants in this way myself. But after I attended a conference of the Society of Plant Neurobiology in the High Tatras of Slovakia the following year, I became nothing short of obsessed with the idea. It was the beginning of a long journey which took me around the world, from the botanical gardens of London, Edinburgh and New York, to India, China, Brazil, Chile and Australia, even the jungles of Mauritius. But the physical distance I have traversed has hardly compared to the mental ground I have covered.

One thing I have come to realise through this work is just how irresistible humans find it to draw big conclusions about the world from individual experience. It's part of what makes us the *sapient* creatures that we are. And what makes us incredibly blinkered.

Even the finest thinkers in human history have been prone to navel-gazing. The ancient Greek philosophers, whose work widely fertilised our intellectual history, saw a world that mirrored their perspective in quite a literal sense. For the Greeks, the centre of Hellenic power—Delphi—was also the centre of the geographical world. They called it the *Omphalos*, the world's navel. It was said to have been the meeting point of two identical eagles that had been released by Zeus from either end of the world. The Delphic Oracle who took residence there was revered across the ancient world. Pilgrims would walk for days to reach the sanctuary in the foothills of Mount Parnassos, because to consult the Delphic Oracle was to tug directly on the cosmological umbilical cord.

I found myself travelling to Delphi in 2019 to join a gathering of diverse minds, including philosophers, scientists and creatives. We were meeting to discuss humanity's place in the world. Whether through earnestness or irony, we met at the navel of the classical world to consider humanity's habit of navel-gazing—and to work out how to move beyond it. The ancient Greeks were not the only civilisation to fall into "Omphalos syndrome," the belief that one's own socio-political centre is the centre of the cosmos. It's been a habit throughout history: as individuals and societies, we all have a tendency to think the world revolves around us. And it has landed us in a great deal of trouble—ecologically, politically and psychologically. Now, this intrepid band of thinkers had met at Delphi to untangle the nature of humanity and our interactions with the environment. To seek out new ways of thinking for a different kind of future—one that might give us a more mature and connected kind of congress with other living things.

During the weekend, we had the opportunity to explore the

archaeological site. As I stood on the forecourt ruins of the Temple of Apollo, surrounded by the brown scree slopes of the mountain, I thought of the two words fabled to have been inscribed there: "Know Thyself." A simple injunction, but a lifetime's work for the individual. Certainly more than a conference's worth even for a hundred intellectuals. I had a strong inkling that we needed to think *very* differently to get deeper into these problems, to learn from other species and come at investigating our own minds in a new way. But I did not realise the full extent of how radical my focus would become.

Delphi was a kind of conversion experience for me. The landscape itself mirrored the problem we were trying to solve: it was filled with history interlaced with the living present, archaeological sites couched in resinous forests and meadows. But we tend to see only the rubble remains and faint imprints of the past. We are only dimly aware of the commerce of organisms for which these human productions are now a stage. It was there that I realised clearly that to "know thyself," one had to think well *beyond* oneself, or even one's species. One can only know *thyself* by knowing *others*. We have to think into the experiences of other organisms dramatically different from ourselves, however rudimentary or complex they might be. So different in fact that their experiences might be generated without any of the familiar animal thinking machinery. No brains, neurones or synapses. I began to think about the *sapience* of plants.

We are so entrenched in the dogma of neuronal intelligence, brain-centric consciousness, that we find it difficult to imagine alternative kinds of internal experience. The title of this book alone might evoke derision and consternation from some. This is understandable: it challenges the foundations of human experience. To begin building a picture of how thinking without brains might be possible, this book will skirt the frontiers of neuroscience, plant physiology, psychology and philosophy, to delve into what it might be like to *be* a plant. I will take the seeds of scientific evidence and cautiously see where they might grow with further investigation.

Caution is necessary: whether you are deeply sceptical of the possibility that plants might have intelligence or are an enthusiastic believer in the supernatural wisdom of other lifeforms, we all need to broaden our minds carefully. To dramatically shift our understanding of the world in a measured way, based on the evidence as it emerges. I neither want to narrow-mindedly ignore the astounding possibilities of what science is uncovering nor to start a new animistic cult of nature worship. This book is written for everyone, both those who believe that plants might be intelligent and those who believe they could not possibly be. What you read here will be a challenge to anyone's preconceptions. So try to let them go, begin with an open mind, and follow the path that the evidence is building for us—if we can allow ourselves to see it.

What we might find may scare us: understanding other ways to be in the world will probably show us that human intelligence is not quite as special as we like to think. We are just about beginning to acknowledge that non-human animals might have intelligence, but accepting that plants might requires a *radical* shift. Losing our assumed place at the top of some imagined hierarchy might seem galling but the rewards of shifting our perceptions will be wondrous. The question is, to borrow from the Dutch primatologist Frans de Waal—are we smart enough to know how smart plants are? I might also add—are we brave enough?

The work begins in our own minds. One of the most powerful tools that Charles Darwin used as he developed his theory of evolution by natural selection was not a scientific instrument or a specimen. It was the motion of his own body through space. Every day, once in the morning and once in the afternoon, he would walk along the Sand Walk, a gravel path bordering the grounds of his house at Downe in Kent. He called this route his "thinking path." In the rain, sun or sleet, Darwin mused over his readings, correspondence and experiments in the passing company of plants and animals. He was one of many thinkers to use the power of physical motion to move the mind forward and help thoughts to grow.

I had hoped to travel to Down House for the final leg of the journey in writing this book, to feel the crunch of the Sand Walk's gravel under my shoes just as Darwin did. I wanted to pen this opening piece among the same privet hedges and trees that leaned in to hear Darwin's own careful, expansive thoughts. Sadly, the obstacles of Covid-19 prevented me from making this pilgrimage in person. In its stead, I mentally retraced the steps of my own "thinking path," the one that I have travelled while seeking to understand plant intelligence over the past two decades. It has been a long and fertile route which has lit up my imagination and opened my mind. I invite you to join me on the journey.

PUTTING PLANTS TO SLEEP

It's not every day that you get to perform a scientific trick in front of a large crowd that *truly* surprises them. On 9 August 2019, in a lecture hall in Mauritius, I managed to shock just such an audience using little more than a glass bell jar, a cotton pad and a small quantity of anaesthetic. My drug of choice was one which veterinary surgeons use on horses, cats and dogs to make them temporarily and safely unconscious. Many people in the audience had probably taken a pet to the vet at some point, seen it slide gently into sleep, but they had never seen a demonstration like this before.

It was the perfect setting for something curious and apparently impossible to happen. Mauritius is one of a group of Indian Ocean islands that, as a result of their isolation, were once full of wonderfully bizarre plants and animals. They are just close enough to mainland Africa and the island of Madagascar for an eclectic cast of species to have made the journey over, but sufficiently far away that these creatures spun off on their own, strange evolutionary adventures once they settled in. The results include the roaming giant tortoises, boucle d'oreille shrubs with blood-red flowers, burrowing boas, wispy Fleur de Lys and, of course, the enigmatic dodo. Since Europeans arrived on the previously uninhabited island at the end of the sixteenth century, many of these species have been lost or imperilled. I had made the trip there for several reasons. The first was an invitation to talk at a special meeting organised by the Institut Bon Pasteur.* The second was to search for the eighteen species of wild vine that grow only in Mauritius, to use for my research at the Minimal Intelligence Laboratory (MINT Lab) in Murcia in Spain. These vines have not been meddled with as domestic species have been; they are wild inhabitants of the tiny areas that remain of Mauritius's once-sprawling natural forests.† For me, they had irresistible experimental potential, so much so that I was willing to travel halfway across the globe to find them.

My talk was in the evening, so earlier that day I had gone vine-hunting with Jean-Claude Sevathian, an expert caretaker of the island's rare plants. Several subspecies of the island's plants even bear his name. From a moving jeep, his eyes could pick out the vines' sinuous forms from the dense rainforest foliage with the most incredible accuracy. Some of the species we searched for were only found in Mauritius's most remote, densely forested reserves, so we were venturing into territory rarely explored by humans. As we sped through the bush, I couldn't help but think of a young Charles Darwin seeking out plant specimens in little-known island regions, though he had reached his by ship rather than via the expediency of air travel. As we scoured the thick green foliage, I imagined

---

* Institut Bon Pasteur (IBP) is a private company whose unique venture is the GEM training and service centre for Geographic Medicine, with which Minimal Intelligence Laboratory was developing a collaboration. The director is Zoë Rozar, my host in Mauritius.
† Only about 2 per cent of Mauritius's healthy native forests remain, most in remote and less accessible regions of the island and offshore islets.

him looking for the first time at species he had never imagined existed. Darwin viewed plants and animals as integral parts of their environment, inextricably woven into the tapestry of relationships with the organisms around them. For him, an animal or plant could only be understood when viewed within this network. A specimen abstracted to sterile laboratory surroundings gave only a partial picture. If we could see life even a little more in the way that Darwin did, our experience of it would be far richer.

I had a third agenda for these explorations, too. I was on the lookout for a suitable patient for my anaesthesia demonstration. I needed one that might be familiar to the audience, could be easily enclosed in my bell jar, and that would be sensitive to anaesthetic. In one of the parks studded with the humped backs of giant Mauritian tortoises, I found a few perfect subjects. They appeared fairly shy and recoiled when touched, but I left them alone for the afternoon to give them a chance to relax.

That evening, I introduced myself to the audience and told them what I was planning to do to the organism sitting on the table next to me. I smiled to myself at the mix of quizzical and sceptical faces that looked back. I made sure that they could all see as I brushed the patient lightly and it folded itself up as it had done in the forest. Then I took a cotton pad soaked in a carefully measured volume of anaesthetic, placed it down next to the subject and lowered the large glass bell jar over both. The bell jar was less for a retro flourish, or to stop the subject from escaping; I needed to steep the air inside with the anaesthetic. I could not use a gas mask for delivery under these circumstances, as a vet might with a dog.

I knew that the anaesthetic would take a while to work, having practised the process several times at my lab to make sure that I got the timing and quantities exactly right. While I went on with my talk, I saw pairs of eyes from the audience darting between me and the bell jar, scouring it for signs that the anaesthetic was working. Just under an hour later, it was time for the big reveal. I called for a volunteer to see if they could try to wake my subject, selecting a woman from a forest of raised hands. She stood up, unfurling her strikingly tall, slender form and walked over. I raised the bell jar up so that she could stroke the subject lightly with a finger, clearly expecting it to recoil as it had done earlier. But nothing happened, even when she tried again. The subject was fully anaesthetised. The audience was silent for a few moments before shocked exclamations and clapping broke out across the auditorium.

Now, this might seem a very strange thing to be surprised about. I wonder if you have guessed the nature of my subject that evening. It was certainly not a mammal, nor was it any other animal. In fact, it was a plant, a *Mimosa pudica* to be precise. The "sensitive plant" is an import from the Americas that now grows wild all over Mauritius. Mimosa is familiar to many people because of its enchanting "shyness": it draws its leaves against its stems as soon as it is touched. This is not just amusing to humans, it is an effective measure against plant eaters, making the leaves tricky for herbivores to get a hold of. Of course, the plant is not really "shy" as we imagine it; this folding is a clever evolutionary trick to stop it getting eaten when it senses something which might be a predator nearby.[1] The anaesthetic took this response entirely away, and the plant remained passive to our touch, much to my audience's surprise.

Some months later, I did the same trick under less formal circumstances, in a classic eighties bar, Planta Baja in Granada, Spain. I was at an event filled with live music and talks called Psychobeers, held at regular intervals by graduate students from the University of Granada. After the acoustic pop band Cosas que hacen Bum played a song, very fittingly called *"Sin prisa, un jardín"* (No rush, a garden), I went over to my equipment, already set up on the stage overlooking the atmospheric bustle. This time I was using one of the ferocious carnivores of the plant world, a Venus flytrap (*Dionaea muscipula*). These plants have specialised leaves which snap shut on any unsuspecting insect that wanders across them. They then exude enzymes into the cavity to digest the body.[2] Many people will be familiar with the fascination of triggering these traps, which look rather like grinning, spike-toothed mouths. The reaction to the plant's movement was nothing compared to the audience's reaction when I rendered it anaesthetised, however. This time I had rigged the whole thing up with a camera, so even people having a drink at the bar could watch what was happening clearly on a screen. I had also arranged surface electrodes to measure the electrical activity in the trap's excitable cell membranes. At the beginning of the talk, the electrical signal showed spikes of voltage every time I touched it, a clear sign of the plant's active interior life, like an ECG signalling the heartbeat of a human hospital patient. After an hour, I asked a volunteer to stroke the Venus's traps. The plant remained totally still. The screen showed a flatline: the spikes of electrical activity that had appeared when it was touched before the anaesthesia were gone.

You might be wondering how exactly the anaesthetic renders these plants so unresponsive. This will be a story for a later chapter about plants' invisible electrical activities and all the ways in which plants use the complex networks of information sent rapidly through their bodies. For now, let's focus on the fact that these abilities can be taken away with the very same anaesthetic that might put a cat to sleep—or you or me, for that matter. It's not just mimosa leaves or Venus flytraps that lose their dramatic abilities under anaesthetic. All plants will stop whatever they were doing when under the influence, whether that be turning their leaves, bending their stems or photosynthesising. Seeds will even halt their germination.[3] In short, anaesthetic causes plants to stop responding to the environment in all the ways they usually do.

This similarity is surprising, seeing as the lineages that produced animals and plants diverged over one and a half *billion* years ago.[4] We are in entirely different kingdoms, and yet can be "knocked out" by the same drugs. To put this into context, even bacteria can be anaesthetised. These organisms are not even in the same domain as us, the highest level of division in the tree of life.[5] Yet these single-celled organisms, like the cells of our bodies and those of plants, are sensitive in just the same way to being temporarily shut down. Even the structures inside our own cells that release energy—mitochondria—and the photosynthesising chloroplasts inside plant cells are sensitive to anaesthetics. To be alive is to be susceptible to anaesthesia.[6]

It might be more accurate to say that we can be knocked out by the same drugs that put plants under, for plants actually create these chemicals for themselves. When we put a mammal to sleep temporarily, we give it a dose of synthetic anaesthetic. But plants synthesise a wide array of such drugs. These substances are released at points of stress: when a plant is wounded, for example, it will release anaesthetic chemicals such as ethylene in its tissues. When a root becomes dehydrated it releases the three anaesthetics ethanol, ethylene and divinyl ether.[7] Why they do this, we don't quite know just yet. Some help the plant activate defence measures while the purposes of others are less clear. Perhaps, like a human going for a pint to relax after a busy day, they are just taking the edge off. Some of these substances are released in such enormous quantities that they even affect the Earth's atmosphere.[8]

Humans have used some of these chemicals for a very long time: the leaves of coca plants were chewed for their anaesthetic properties for thousands of years before cocaine was isolated and became the first local anaesthetic and then a recreational drug. You can find thymol from thyme leaves in your mouthwash and eugenol from clove oil is used as a local dental anaesthetic.[9] This is not to mention the vast array of other substances produced by plants that we intentionally use to affect our minds and bodies: tobacco, ethanol, aspirin, marijuana, caffeine-laden tea leaves and coffee beans. Many medicines that we use today originated from plants, either extracted from plants or based on bioactive chemicals produced by plants. They include the antimalarial quinine from the South American tree, *Cinchona officinalis*, and digitoxin used to treat heart failure, extracted from *Digitalis purpurea* or common foxglove. We might be distant from plants in evolutionary terms, but we are still intimately involved with them through many biochemical cross-links.[10]

Experiments with anaesthetics are not only surprising from an evolutionary standpoint. They provide the perfect blank slate from which to begin to view plants in an entirely new way. If we can reduce them to anaesthetised bodies, like a pet ready for surgery, we can begin to become more aware of what plants are like when they are fully functional. From the outside, a plant under anaesthetic stops "doing" the things that it is usually busy doing. When the drug wears off, the plant resumes these activities, after taking a little time to reposition its leaves and compose itself. In the case of the Venus flytrap, if you touch a trap when it is first recovering from the drug it will close, but only *very* slowly.[11]

We might then refer to the things it is normally doing as the plant's normal *behaviour*.[12] Are plants usually *behaving*? This might seem like a strange word to use in relation to plants: it goes against everything we intuitively assume about them as inert, passive organisms, rooted in the soil. The definition of "behaviour" from *The Penguin Dictionary of Psychology* is a useful reference point:

> A generic term covering acts, activities, responses, reactions, movements, processes, operations, etc.; in short, any measurable response of an organism.

We tend to see plants as the background foliage to the rapid comings and goings of animal activities. But can a mimosa folding its leaves or the Venus traps shutting not at the very least be defined as reactions, movements and "measurable responses," terms we might use to describe animal behaviour?[13] Perhaps the parallel effect of anaesthetic drugs on a plant, a cat or a person might give us pause to rethink our prejudices.

Now we are faced with an important question: what does it mean when you take away mimosa's ability to fold its leaves or disarm the Venus traps? Beyond stopping it from moving or responding, what does it mean to *put a plant to sleep*? We know what being anaesthetised means subjectively in the case of an animal or a person: a state of awareness is removed, we are being rendered from a conscious to an unconscious state (a shift which the uncharitable reader might reserve only for humans). The word *anaesthesia* itself has its origins in the Greek word *anaisthēsia*, meaning "insensibility" or "inability to perceive."[14] In your brain, this means that the electrical activity in its cells is compromised, just as in the Venus flytrap I anaesthetised. They no longer respond to stimuli. The exciting—and controversial—implication is: if a plant can be temporarily put to sleep, as an animal can, does that mean it also has some kind of "waking" state normally? Perhaps we might consider the possibility that plants are not simple automatons or inert, photosynthetic machines. We might begin to imagine that plants have some kind of individual experience of the world. They might be *aware*.

If plants are aware, then they must have some kind of interchange between their internal state and the external environment. They must be able to collect information from the outside, process it and use it in more sophisticated ways than simply *reacting* to things. What if plants could store information and use it to make predictions, learn and even plan ahead? We are just starting to discover instances where plants might be doing this, but they are complex feats to get to the bottom of. In the following chapters, we will explore the exciting clues that cutting-edge research is revealing as to what plants might really be experiencing and what they are really doing. We will gather these together into a radical new picture of plants as organisms that are not only aware but highly engaged with the world.

We can begin with a simple example, an unassuming little flower called Cornish mallow or Cretan hollyhock—*Lavatera cretica* to the botanists. It has a penchant for alpine regions in the warmer climes of southern Europe and North Africa, but can often be found as a domesticated expat in the gardens of cooler countries.

Many plants are "heliotropic," meaning they follow the movement of the sun through the day.* You might have seen dramatic time-lapse videos of fields of young sunflowers turning their tips dutifully to follow the sun across the sky. We will meet these plants and their surprising abilities properly in a later chapter. For now, let's give the humble little *Lavatera* a moment of attention. It is also a sun-worshipper, but a well-prepared one. Throughout the day, its leaves turn to face the sun. This maximises the amount of light they soak up, rather like vacationing humans shifting their beach chairs to escape encroaching shadows. During the night, though, *Lavatera* turns its leaves to face the sunrise before the sun is even up. This doesn't simply mean that the leaves spring back into the position where they started at the beginning of the previous day. More astounding still, it can hold information about what direction the sun will first appear from for several days, even in the total absence of any sunlight. *Lavatera* plants kept in darkness in the lab will accurately predict the direction of sunrise, dutifully turning their leaves to face the absent sun each night. Only after about three or four days do they lose the plot a little (as most of us might).[15]

The timing of these leaf movements is controlled by the cycle that binds organisms to the daily cycles of day and night, the circadian rhythm. This is another of those universals of living things, another biochemical link that we share with even our very distant relatives on the tree of life—from plants and animals to bacteria.[16] We know that our own daily circadian rhythms are controlled in part by the production of a chemical called melatonin. The levels of this hormone increase and decrease at different times in a 24-hour cycle and control how awake or

---

* This is the Latinate term for "solar-tracking" that botanist Augustin Pyramus de Candolle coined in the early nineteenth century.

sleepy we feel, as well as myriad other processes in our bodies, from metabolism to our body temperature. It is produced in the pineal gland, a minute organ in the centre of the brain which has acted as a kind of light receptor throughout animal evolutionary history. The French philosopher René Descartes called it "the seat of the soul," the originator of thought and action.[17]

Oscillations in melatonin levels allow an organism to predict what state it should be in at any one point. If it had to rely purely on reacting to its environment there would be unhelpful delays, such as being awake for a period after the sun was down, or being inordinately slow to get moving in the morning (though some of us still might have this problem). You might have taken melatonin pills to counteract the effects of jet lag, overriding your own internal melatonin synthesis to retrain your system into a new time zone. We will see later how plants, too, can experience a form of jet lag if manipulated in the lab. Plants also make their own melatonin, phytomelatonin.[18] It was only named in 2004, several decades after melatonin was first discovered, because it was assumed that only animals produced this chemical. They have circadian rhythms that control their inner workings as well, including *Lavatera*'s nocturnal movements. Plants' state of "awakeness" is altered on a daily basis, and with minute precision,[19] by their own internal rhythms, not just by the dramatic effects of anaesthesia.

We must open our eyes to entirely *different* ways of doing complex things. *Lavatera* manages to do something that appears to be strikingly smart. It may be nothing more than an ingeniously evolved trick, but even if it is, it points to further underlying complexities. It could point to something like *intelligence*. There is no one single, agreed-upon definition of what "intelligence" is. Drawing analogies between what plants such as *Lavatera* do and our own capacities is unavoidably risky, which is why understanding plants better has the potential to show us an enormous amount about how our own minds work.[20] For now, let us just sow the seed of the idea that intelligence has something to do with the nerve-like processing of information. What *Lavatera* and other plants manage, they do without using anything we might think of as a "brain." We currently have a very narrow view of what it takes to be intelligent, automatically writing off anything without a recognisable brain, or at least a well-developed hub of neurones. We used to assume that intelligence must have evolved from one branch in the tree of life along with a certain type of brain. But this picture has been shattered by our recently increased understanding of creatures such as octopuses, which have multiple brains in different limbs and astounding mental capabilities. We need to rethink our understanding not only of whether other organisms, including plants, might be intelligent, but what intelligence *is*.

This begs another question: do we need to rethink where intelligence can reside? Perhaps intelligence is not something that can only be produced by vast assemblies of animal neurones. It might be possible to produce intelligence from very different kinds of systems. Plants, including our mimosa, use electrical signals like the action potentials that fire along our neurones, use ion movements and have cells that can transmit them relatively long distances through their bodies. Looking at an analogy, comparing the ways in which animals and plants move, will help to frame the question. Motor information is transmitted to contractile cells in animal muscles, which then execute the movement. In plants, information can be transmitted through specialised fibres with contractile properties in motor organs. This plant motion system operates in an entirely different manner to that in animals. But perhaps some fibres may be considered as "plant muscles."[21] They have a great deal of similarity in function to animal muscles. Perhaps we should not arbitrarily separate them just because they are made of different tissues and operate differently. So, to bring our focus back to less concrete functions: if plants "think" using different systems than animals, does that mean that they are not "thinking" at all? Surely, we should be more open-minded in the way we see organisms built from largely different blueprints. This question is what we are going to explore as we forage deeper into the plants' world.

We could even ask, why *wouldn't* plants be intelligent, as animals are? Animals and plants have evolved intelligence separately, helping them to function in very different ecological situations. On the one hand, we have an animal intelligence that helps us operate as mobile, quick-moving creatures with bodies that always grow roughly the same way. Plants, on the other, have to make it in life as rooted, slow-moving organisms that have to grow creatively instead of just walking off. In order to survive, they need to integrate many different sources of important information—about light quality and direction, which way is up and whether there is something or someone in the way—and use it to control their patterns of growth and development. Plants are constantly, and tirelessly, swaying their organs, responding to uncertainties such as soil structure, predators or competitive neighbours. Plants have to plan ahead to achieve goals. They are not merely passive organisms taking life as it comes, while doing photosynthesis. They proactively engage with their surroundings. Like animals in the bloodied tooth-and-claw wilds, plants couldn't afford to do otherwise.[22] We will delve into the internal experiences of plants, as far as we are able, to understand how they perceive and deal with the complexities of their surroundings.

Intelligence is an elusive quality to perceive in organisms so very different from ourselves, and requires some very clever experimenting. Understanding the possibility that it might exist in completely different forms also requires the kind of open-minded observation that Darwin advocated. That was one of the central goals of my trip to Mauritius. From my work so far, it has become very clear that there are some dramatic differences between domesticated vines and those that live in the wild. The coddled domestics, always provided with supports to climb up, fertiliser, aerated soil and adequate space, have been softened. They are the spoiled lapdogs of the plant world, trained to survive only in sanitised human environments, without competition or hardship. They would not last long out in the forest. The wild vines, on the other hand, have the hardened street-smarts of Mafia bosses with well-established networks of allies and enemies. They have fought fiercely for everything: light, rooting space, climbing supports, to protect their leaves from being eaten. They know who they can work with and trust to cooperate with them.[23]

If we want to find plant intelligence, whatever form it takes, we need to look to the survival-sharpened wits of plants in the wild—not with the eyes of plant scientists used to seeing domesticated crop plants in the lab, but with the astute eyes and open minds of naturalists. To help us see in this more holistic way, to answer the many questions that a revolutionary take on plants will raise over the coming chapters, we will call on many areas of scientific research, but also on other areas of thought, such as philosophy. We cannot radically shift our understanding and perceptions if we limit ourselves to orthodox scientific gospel. We must draw on many different tools for enquiry, to cautiously strike out into the unknown. *Planta Sapiens* will therefore be a confluence of many bodies of thought with deep roots, which will entwine together to grow into new spaces.

Understanding plants in a new way could dramatically change the way we see the world. I know from long experience and the many debates that I have had with my colleagues in other areas of science that the ideas we will explore in *Planta Sapiens* are at odds with most people's perceptions of plants.

They might even make you a little uncomfortable, or force you to wonder what words like "behaving" or "awareness" can possibly mean for a plant, never mind "intelligence." You are not unusual. It is entirely normal, as an animal, to have reservations about applying to rooted photosynthetic organisms ideas that we normally apply only to mobile, animal-like creatures. Most people are probably more comfortable describing the behaviour of an amoeba than of a vine, or the awareness of a woodlouse than a sunflower. You would probably be perfectly happy thinking about a jay burying acorns as "planning ahead," while a plant "planning for the future" might make you feel a little uneasy. We will look at the many sources of your discomfort in the next chapter, exploring the numerous zoocentric traps that limit your perception and the long history of animal-focused indoctrination that has shaped your ideas. By fleshing these out, we will be able to unpick them, and hopefully pave the way for what is to come.

PART I

SEEING PLANTS ANEW

To see takes time.

Georgia O'Keeffe

CHAPTER ONE

PLANT BLINDNESS

There is a problem that afflicts us all from a very young age. It inhibits the way we see the world, but most of us never even know that we suffer from it. We might think that we are aware of our surroundings, that we notice the details of our environment. But we are more often than not floating around in our own personal bubbles, through which only a very small part of the things we see, hear, touch and smell filter into our conscious awareness. The late-nineteenth-century American psychologist William James wrote:

> Millions of items ... are present to my senses which never properly enter into my experience. Why? Because they have no interest for me. My experience is what I agree to attend to ... Each of us literally chooses, by his ways of attending to things, what sort of a universe he shall appear to himself to inhabit.[1]

For most of us, this personal universe is an animal one, filled with rapid comings and goings, especially the electric social hum of human existence. We all but ignore the photosynthetic creatures that make up much of our environment. Most of us, we could say, are "plant blind." We can see plants, of course, but we don't *notice* them, except if they are doing something spectacular with their flowers, or getting irritatingly entangled with our bedding plants. There are some very good reasons for this, which we will explore, but there is also a great loss in giving in to such an inclination. And, if we can work out how to transcend it, we might appreciate the world around us significantly more.

It is hard to understand quite how profoundly limiting plant blindness is without seeing it in action. Every year I give a talk to students in late secondary school. I like to play a game: I show them a series of the winning images from the Wildlife Photographer of the Year competition, which exhibits at the Natural History Museum in London annually. I ask if they notice anything strange about these pictures. They often pick up on some detail of an image, a bloodthirsty bird or an insect carrying an impossibly large object. In all the years I have been doing it, they have always missed the strangest thing of all. There are photos of "Animals in their Environment" and "Animal Portraits," categories for interesting behaviours for "Amphibians and Reptiles," "Mammals," "Birds" and "Invertebrates." Then there is the "Plants and Fungi" category. Have you noticed anything odd? The animals, which make up a tiny proportion of the species on Earth, are attended to from all angles.[2] The plants and fungi, two entirely different kingdoms on the tree of life, are lumped together into one entry. Not one student has ever noticed this.

The same problem is rife among even my own undergraduate students at the University of Murcia. I asked them to estimate how many plant species there were in the carefully curated botanical gardens scattered throughout the campus, which they pass through every day. Most said about ten, a brave few as many as forty. In fact, there are over five hundred species of plants from a vast array of families and habitats.[3] Plant blindness starts early and only gets worse as we let it set in.

There are fundamental differences between our attention to animals and our attention to plants, and these are deeply embedded in our visual systems. This is a tricky phenomenon to model and quantify. One study used a core tool from visual cognition studies called "attentional blink."[4] "Blink" is when the focus that is given to one object slows down our ability to engage with a new object. Our visual processing power is a finite resource, so the more attention the first object takes up, the slower we are to shift on to the second. In this study, one group of people were first shown an animal and another group were first shown a plant. A second object, a water droplet, followed in quick succession. Those looking at an animal initially were much less likely to see the water than those first looking at a plant. The plant simply took up less of their attention, freeing up capacity to notice other things. Plants are not only *thought of* as less interesting, they are fundamentally given less processing power in our visual system, becoming a mass of crowded, static background greenery. The root cause of plant blindness runs deep.

At one level, this isn't surprising. We cannot possibly take in every piece of information available in our environment, our brains would be completely overloaded. We have to filter out the things unimportant to us. Our senses and brains are very good at doing this without us even noticing. One recent calculation estimated that our eyes generate over ten million bits of data per second, out of which the brain processes only sixteen bits in active awareness. Just 0.00016 per cent of the data our eyes create is actually used by the conscious mind (though more, of course, may affect us subliminally).[5] The nature of this filtering has been shaped by our evolutionary history, the kinds of problems that faced our ancestors. If you think of what the salient information would have been for most hominins in the past, spotting predators or seeing animal game spring to mind. Plants have been important, but never quite as *immediately* so: they aren't going anywhere, and they aren't about to attack us.[6] Our eyes and minds have developed to focus on the quick-fire problem of animal movements and forms.

The term "plant blindness" was first coined in the 1990s by biology educator James Wandersee and botanist Elizabeth Schussler. They surveyed nearly three hundred US school children of different ages and found that very few had any scientific interest in plants, especially the boys. This was, they argued, not only because of "zoochauvinism" or zoocentric attitudes among US youth and their educators. Wider society in the West suffers from an inability to see the unique beauty and biological features of plants, to notice plants and recognise their ecological importance and economic value to humans.[7] Even the majority of scientists, who might be expected to have a somewhat more objective view of things, largely see plants as only the inferior backdrop to the animals they want to study. All despite the fact that plants form the basis for most ecosystems on the planet. They also make up one in eight species threatened by extinction.[8]

As the "attentional blink" experiment shows, the problem of plant blindness is fundamental. Growing up, children take far longer to recognise that plants are alive than they do other humans and animals; it's only by about the age of ten that they have come to see apparently inanimate plants as living beings in their own right.[9] This prejudice against plants is hardwired into us, then reinforced by how we are taught to engage with the world. We cannot change our hardware, but we can change how we think about plants collectively, and the way we direct our attention. As William James described, we can *agree* to attend to plants. When plants make it impossible to ignore them, we do attend. If they are capable of stinging us or poisoning us, or offer up vibrant signs of edible offerings, specific plants can become very prominent points of focus. The innocuous-looking leaves of poison ivy can become instantly recognisable to anyone who hikes in North America, and the ripe fruit of blackberry bushes is hard to miss for foragers. If we can make plants easier to observe, our attentions will naturally follow. One study showed that when school children made their own time-lapse videos of plants, speeding them up to animal-like timeframes, they became more interested in learning about them.[10] Perhaps, if we can focus on this dormant awareness, develop new cultures of seeing, we can start to wake up and become attuned to a green world.[11] We might become able to perceive the intelligence of different kinds of living things, not only the ones with brains.

The Great Chain of Being

Our minds are shackled both by the limitations of our senses and by our history. Before Darwin's work unfurled the organic world on a branching evolutionary tree of life in the nineteenth century, living things were ordered in a long, vertical hierarchy. At the very top were God and his angels, and from there cascaded a chain of creatures from Man down to large animals, then rodents and those thought to spring spontaneously from inorganic matter, insects and amphibians. Right at the bottom were the things that did not move, the plants, the bedrock of life. Along with corals and sponges, they were only one rung above the inorganic things such as minerals. This was the Great Chain of Being, which tied all things in the world into a system of value, from lowest to highest. And this value was very much predicated on animal qualities, especially how much something reflected humanity, the pinnacle of theological perfection. This was the dominant view of the natural world in the West for hundreds of years, far longer than our understanding of the evolutionary relationships between things has existed.[12]

The Great Chain of Being still permeates our intuitive under-standing of other organisms. How like us are they? We still place things on a scale of importance from unicellular to multicellular, simple to complex, invertebrate to vertebrate, "instinctual" to "intelligent." Even well-known scientists firmly entrenched in evolutionary theory are still wrapped up in the Great Chain of Being. James J. Gibson, a renowned ecological psychologist of the twentieth century, was oblivious to plants' capacities. He argued that:

> The environment of plants, organisms that lack sense organs and muscles, is not relevant in the study of perception and behavior. We shall treat the vegetation of the world as animals do, as if it were lumped together with the inorganic minerals of the world, with the physical, chemical, and geological envi-ronment. Plants in general are not animate; they do not move about, they do not behave, they lack a nervous system, and they do not have sensations.[13]

Like medieval theologians, Gibson lumped plants together with inanimate rocks. Not only that, he assumed that other animal species also perceive them in this way. Ironically, Gibson's work actually furnishes us with the best framework we have for understanding plant intelligence, as we shall see later. But the fundamental attitude that still permeates science is that plants are verging on inert. The problem with this perspective is, we are only one small part of a kaleidoscopic variety of ways of being alive. Seeing things through the lens of the Great Chain blinds us to much of the biological wonder around us, the con-nections of organisms within an *ecosystem*. Evolution has not produced a linear string of creatures from simple to complex, it has not produced a hierarchy in which intelligence bloomed on the top rungs. Each species is shaped by the pressures of its particular environment and lifestyle, in a vast branching delta of lifeforms. Sometimes this means staying apparently still or simple. Sometimes it means evolving sophisticated, alternative ways of being whose complexity is invisible to us with our anthropocentric outlooks.

This state of affairs is not a given. While our senses might be geared to attend to animals rather than plants, the cultural blindness that afflicts us is widespread but specific; it is not universal. Plenty of human societies in other places and times have overcome the predilection of our sensory systems for rapid movement and distinct colours. Animistic societies in pre-Christian Europe or in different parts of the world today have viewed plants in a very different way, as entities with potency and meaning.[14] In certain cultures, such as the Maori or some American Indian groups, plants are seen as kin with a shared heritage. In Amazonian cultures, as well as among the Inuit and the indigenous peoples of subarctic Canada, plants, like animals, are seen as "persons" with souls of equal standing. They can be part of social interactions, just in the same way that people and a few, privileged animals can be in the blinkered West.[15] We needn't believe in souls to change the way we value and understand other life. If we can shake up the scientific orthodoxy, use our powerful scientific tools to investigate in a more open-minded way, we might find all the evidence we need to see that plants are far from just the substrate for animal life.

Mobile minds

How did we get this way? The repercussions of the Great Chain of Being affect us all still, making us assume that intelligence belongs to things with animal-like properties—moving freely, feeding on other organisms, having sex or communicating with one another. But these are misleading proxies for intelligence, founded on historical prejudice. Philosopher of neuroscience Patricia Churchland, in her 2002 book *Brain-wise Studies in Neurophilosophy*, insists that:

> first and foremost, animals are in the moving business; they feed, flee, fight, and reproduce by moving their body parts in accord with bodily needs. This *modus vivendi* is strikingly different from that of plants, which take life as it comes.[16]

Churchland echoes the general consensus that animal movement requires intelligence, while plants stay rootedly and stupidly put. This is a misapprehension on several counts. Plenty of single-celled organisms more distantly related to us than plants are overactive busybodies and plenty of animal species tie themselves to one spot, temporarily or permanently. Take the corals for example. These tiny animals build calcium carbonate homes around themselves on shallow, brightly lit seabeds. They clone themselves year after year, until they have created limestone palaces which are the foundation of sparkling reef worlds supporting about 25 per cent of all marine species. Corals have colourful live-in tenants, called zooxanthellae, which harvest sunlight and make food for the corals, much as chloroplasts do in plants.[17] The larvae of these corals are minute and mobile, tossed on the waves until they find a suitable spot to up-end and settle down permanently. It's much the same story for sponges and an array of other marine animals such as mussels and clams. Yet many people are not aware that corals are alive, never mind that they are animals. They are often mistaken for something plant-like.

Are corals smart? Possibly smarter than you might expect for minute, static creatures. They can switch between their diets of sunlight and hunting for prey with tiny tentacles, and they go to war with one another over territory. But their swimming larval stage is their least self-possessed phase.[18] In corals, then, motility does not seem to denote intelligence. It is when corals are sedentary that they engage in those activities, which would seem to contradrict Patricia Churchland's argument that

[i]f you root yourself in the ground, you can afford to be stupid. But if you move, you must have mechanisms for moving, and mechanisms to ensure that the movement is not utterly arbitrary and independent of what is going on outside.[19]

If we can think outside the box, we could turn Churchland's assertion on its head. If you can move freely, you can correct your mistakes. If you are rooted in the ground though, growing and changing the way you are arranged is your primary means of fine-tuning yourself to your environment. This takes time, from minutes to hours to days. Most of the changes plants are capable of making are relatively much slower than the lightning-fast reactions of animals (though, as we saw with the Venus flytrap, plants can be speedy if needed). If plants can't be intelligent and predictive about how they move and grow, they lag behind whatever is going on. In the cut-throat worlds in which wild plants exist, trailing behind means your competitors will overrun you and your predators will eat you up.

On the matter of being rooted, one thing that makes it very difficult to see what plants are doing is that much of their activity happens imperceptibly, underground. We think of plants in terms of their visible parts: shoots, leaves and flowers; the roots are just the nutrient- and water-absorbing anchors. In fact, the roots are unbelievably complex, and can make up over half the biomass of the total organism.[20] They spread the plant's reach far from the main stem, collecting information over space and time about the living and non-living environment to allow the plant to make the most of how it grows and uses its resources. Individual roots can navigate towards useful things like water and minerals as they grow, but also around objects, avoiding obstacles before they even touch them. Plants will extend their root networks in areas where resources are increasing over time, and pull back from places where things seem to be going south, like traders in an underground stock market. Roots create an interconnected signalling web between plants, where unseen messages pass between drought-stressed individuals and those assailed by the attentions of herbivores to allow their neighbours to take pre-emptive action, or even sync their flowering more closely to each other. Roots are the channels by which friend and foe are identified, and subterranean territory wars are waged.[21]

Some researchers argue that it is more accurate to think of the roots as the "head" of the plant and the green parts as the posterior.[22] The root system senses aspects of the environment, like the sense-organ-laden head of an animal, and directs the activities of the rest of the plant's body. Conversely, the shoots and flowers are involved in the baser aspects of the plant's life: absorbing sunlight to produce food—like the animal digestive system; and sex—an analogy too obvious to spell out. If we picture plants as intelligent organisms up-ended into the soil, rather than static clusters of shoots anchored by roots, it might make it a little easier for us to understand them.[23] This invisible network underground is of great frustration for plant scientists, because getting access to the roots by digging them up usually destroys them, so there are many mysteries still to be solved about the so-called "root-brains" of plants, a concept that can be traced back to the work of Darwin.[24] As one tree researcher, Scott Mackay, has put it: "Below-ground is kind of a frontier, an area of research that's becoming more and more important."[25]

Not only that, but the root-brain is a hybrid one. The roots of plants are intimately entangled in complex relationships with another very misunderstood kingdom of organisms: fungi. When you think of a fungus, you probably think of the fruiting

mushroom bodies that you can chop up and put into a stir fry
or that emerge as if magically from rotting logs. You probably
don't imagine the vast network of mycelium strands that per-
meate through the soil and whatever the fungus is feeding on.[26]
These invisible threads are really what fungi are made of. The
largest organism on Earth, in fact, is probably a honey fungus,
Armillaria solidipes.[27] There is one in the Blue Mountains of
Oregon which is about two and a half miles wide at one point,
as far as it is possible to estimate, making a giant redwood, never
mind a blue whale, seem petite by comparison.

Here we come across another paradox of plant blindness: we
tend to think of things with the ability to feed on other organisms, as mammals do, as more intelligent. After all, you have to
outwit your food, right? This kind of diet is called *heterotrophy*,
whereas plants, which produce their own food using the power
of sunlight and photosynthesis, are *autotrophs*. Well, fungi with
their root-like mycelium and ephemeral fruiting bodies are het-
erotrophs just like us. Their mycelia use enzymes to break down
the tissues of other organisms and allow the fungus to absorb
them for food. But most of us would probably balk at the idea
that a fungus had the same kind of diet as an animal, because
most of us wouldn't consider a fungus to be particularly smart.
As with movement, the "animal" quality of heterotrophy is not
a good indicator of intelligence. It doesn't even divide neatly
between these kingdoms: animals such as corals can co-opt pho-
tosynthesisers and plants can be carnivores, as we saw with the
Venus flytrap. We need to look in a different way.

Rooted congress

The delicately gilled forms that emerge silently from fungal
mycelium have one purpose. They are the sex organs of the

fungi, spreading tiny spores on the wind that will sail off to seed new networks of fungal threads elsewhere. These fruiting bodies are the portion that humans notice because they are visible and *sometimes* edible. Likewise, the flowers that many plants produce are seen as aesthetic jewels cultivated for human pleasure. One study from the 1980s found that many children did not even consider an organism to be a "plant" unless it had flowers.[28] We festoon our homes, celebrations and artwork with flowers, and spend inordinate sums on their ephemeral beauty. This rapture has, at some points in history, reached levels bordering on hysteria. During the Dutch "tulip mania" in the early seventeenth century, the highly saturated colour of the tulip flowers and their novelty made them the focus of a spectacularly inflated market. A single bulb could be worth five times the value of a modest house or ten years of salary for a skilled worker, until prices collapsed dramatically in 1637.*

But a strange form of denial has been coupled with this attention to petals. Our love affair with flowers has been a blind obsession. Until the nineteenth century, many scholars vehemently denied that plants were sexual organisms at all. The idea that animals moved and had sex and plants were still and asexual was ancient, traceable to Aristotle, Plato and other classical authorities. In the seventeenth century, a few naturalists—such as John Ray and Nehemiah Grew at the Royal Society of London— did speculate that pollen was a fertilising agent. The German botanist Rudolf Jakob Camerarius even carried out experiments showing that pollen was necessary for seeds to form, and published his research in *De sexu plantarum epistola* (1694).[29] But these novel ideas were not adopted by the mainstream, the old

---

* Though the "mania" was neither as extraordinary nor as ruinous as older accounts suggest, it was a dramatic cultural event. See Goldgar, A. (2007), *Tulipmania: Money, Honor, and Knowledge in the Dutch Golden Age*. Chicago: University of Chicago Press.

classical distinctions remaining the popular opinion. We have always been captivated by plants' reproductive parts, but sex is seen as something that goes along with movement: it is something that animals do.

The group of plants that produce flowers are called the *angiosperms*, from the Greek words meaning "vessel" and "seed." Flowers can contain both: the pollen-producing anthers and the egg-containing ovaries. The thing about sex between rooted organisms is that it happens at a distance. Like parted lovers in a Shakespearean romance, their intercourse requires some kind of intermediary. The corals can release their sperm and eggs into the ocean currents in an orgiastic explosion, which is mysteriously coordinated to a single night. Land plants can use wind or water, but they frequently use animal interlocutors. Rather than seducing each other, plants seduce animals with sensory delights into being their illicit go-betweens. The beauty which attracts us to flowers draws in bees, wasps, hoverflies, hummingbirds, honeyeaters and the innumerable other species of pollinators. These animals are seeking nourishing nectar and pollen, but these are only rewards for the service they provide. Their real role is as unwitting couriers of pollen between photosynthetic lovers.

It is these intense relationships between plants and their pollinators that have led to the vast diversity of the flowering plants. When angiosperms first evolved 130 million years ago, an evolutionary cascade began which changed the entire world.[30] They radiated into so many species that they now outnumber all the non-flowering groups—about 230,000 in all. Flowers are exquisitely specialised: their appearances tailored to particular types of vision; their shapes formed for particular beaks, proboscises or snouts; the quantities of precious nectar titrated carefully depending on the appetites of the go-between. The sensory world of pollinators is something we can only dimly imagine. Hummingbirds, for example, can see a colour spectrum that extends well into the ultraviolet range. They see colours and patterns in flowers that we cannot, which reveal advertisements, landing strips and guide patterns to help them on their quest.

These floral messages might be true, but sometimes, animals can be tricked. Seduction can be deceptive. The bee orchid, for example, will play on the lustful attentions of male bees by mimicking a female orchid bee in appearance and scent. Male bees try to mate, and the unsuspecting suitors are loaded with sacks of pollen, which they then carry to other bee orchids in search of satisfaction. The male bees are simply the playthings of the orchids' congress. If you think that the mobile animals are in charge, you would be wrong. Pollination is an ongoing evolutionary game between animal and plant.

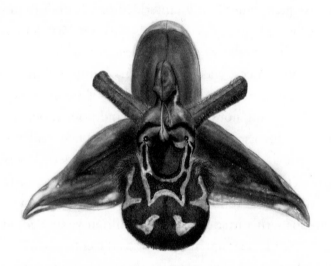

Like the other apes, we have only served plants to spread their seeds as simple fruit eaters; our visual systems do not extend into the fantastical ranges of the ultraviolet or infrared. Angiosperms do include almost all the plant species that humans use for food, though, which we have exploited through selective breeding and agriculture since about 10,000 BC. It is almost unbelievable that, in exploiting plants' sexuality and fecundity, we have still not granted it to them. Writers and artists have used flowers and fruit as coy analogies for fertility and sex, from renderings of biblical scenes by Renaissance master painters to social media influencers today. Think of the paintings of Georgia O'Keeffe for example. Her *Two Calla Lilies on Pink* (1928) is floridly erotic, the tapering pale blooms and their protruding yellow spadices highly suggestive. Yet this painting would never be banned from a social media platform as an explicit painting of human genitalia might. O'Keeffe's sensual floral paintings sit in stark contrast to her images of bleached animal skulls. Flowers and skulls: sex and death. So why do we find it difficult to see flowers literally as organs of plant sex, while freely using them as gentle erotic metaphors?

In the game of seduction between flowers and their pollinators, there is something else going on which we deny non-animal organisms. Plants have been involved in a flirtatious conversation with animals ever since flowers began to evolve. They signal to pollinators that "nectar is here," then shut up shop in the nectar glands when a pollinator has visited—forcing the animal to move the pollen on. Plants and fungi have been in an even longer-term relationship. Fungi have the chemical tools to harvest from the soil precious resources such as phosphorous and nitrogen that are difficult for plants to obtain. Plants have the alchemical ability to create sugars from sunlight through photosynthesis, to which they allow the fungi access. The relationship is mutually beneficial, and has lasted over 450 million years as a result. Fungal threads also form part of the interconnected underground web, along with plant roots, linking the root networks of many neighbouring plants and allowing the traffic of useful materials and important messages. They form what has been called the "wood-wide web."[31] If plants can have such an interchange with other plants and other species, would it not be less of a stretch of the imagination to think that plants might be able to communicate within their own bodies, in a complex way that might be akin to "thinking"? This, as we shall see, will be at the core of seeing plants in a new, proactive light.

Biosemiotics is the idea that all life is involved in *meaning making*. It has been defined as "the study of distinctions that make organisms, what they recognize, what they intend, and what they know."[32] This happens at the level of single-celled organisms, which can collect information and make decisions. The plasmodium of the slime mould *Physarum polycephalum*, for example, is an amoeba-like cell with some surprising abilities. When presented with a maze in the lab, it can find the shortest route through it in a way that would be impossible were it only to be responding to basic environmental signals with behavioural reflexes.[33] You could say that the plasmodium has its own perception of the world, composed of a wide array of information collected from the environment, which it evaluates and uses to make decisions for future behaviour. If relatively simple, single-celled organisms do this, why wouldn't complex, multicellular plants?

We are very comfortable with the idea that plants will grow towards the light. Everyone has seen the stems of a plant set on a windowsill lean enthusiastically towards the window pane, forcing us to rotate the pot if the plant is not to become overly precarious. If we break down the process by which the plant manages this, it becomes less simple than it appears at first glance. The plant must detect the direction of the source of light and communicate this information to its tissues to direct the pattern of growth in its stems. Light changes at a much faster rate than the plant can grow, so there must be something more complex and calculated than a reflex response going on in the plant's growth processes. As we saw in the last chapter, plants can plan ahead with their leaf placement. They can store information about where the sun will come up even when kept in the dark for a few days.

Extend this out to the other kinds of information a plant might collect over time—where water and minerals are in the soil, the approaches of a herbivore, temperature changes, what their neighbours are doing to defend themselves or reproduce, the daily cycles of sunlight—and there begins to emerge a rich tapestry of experience which the plant might use to determine all of its internal activities, bodily movements and growth patterns on a short- and long-term basis. It becomes a sensate being, with its own *Umwelt*, that makes meaning from the world. We, with our animal senses and speeds, find this hard to detect, but we can begin to imagine it. If we privilege animal and especially human channels of communication and ways of knowing, we will miss the bulk of the meaning in the natural world around us.

Background greenery

Our predilection to prioritise animals over plants is deeply ingrained, even in places that should be oases of plant focus. In the spring of 2019 I went to visit the Royal Botanic Gardens at Kew near London. This is the world's most famous botanic garden, home to over 50,000 living plants from all over the world. At the entrance, visitors pass by the Kew Mural, a wooden-relief sculpture depicting the catastrophic storm that killed or damaged more than 1,000 trees on 16 October 1987. It is a stunning piece, constructed carefully from many different kinds of wood collected from trees that were felled by the tempest. The ash, oak, hornbeam, lime, beech, elm and other timbers used create a stunning patchwork landscape of polished colours and grains.

But there is something strange about it as well. I noticed as I looked closely at it that about two-thirds of the sculpture is devoted to images of animals displaced by the storm, two of whom are a pair of Chinese imperial lions from the ornamental gardens, animated into living creatures. The wind itself is personified. Yet the trees and plants, the organisms from whose broken bodies the very mural is made, are little more than background. Even on the gates to this Eden of plant science. Plants underpin much of life on this planet, yet our animal speed makes them invisible to us, even in a place where plants have been central to a rich scientific history. We need to shape narratives that put plants centrally, actively drawing our attention to them, to fully reflect the leading roles they play in our ecosystems and economies.[34] We can start by paying attention: focusing on the details of how plants really behave, dissolving our assumptions that they lead inert, static existences. In the next chapter, I will push open the door to the plants' world, so we can begin to change our perspectives and see theirs more clearly.

CHAPTER TWO

SEEKING A PLANT'S PERSPECTIVE

"I am getting very much amused by my tendrils," Charles Darwin wrote to his friend Joseph Hooker one day during a long, beleaguered summer in 1862. He had been confined to his sickbed for weeks, suffering from a nasty bout of eczema. His only solace had been watching the filigree tendrils of young cucumber plants explore their environment as they grew from their pots on his windowsill. Over the many hours that Darwin watched, they ranged in circling motions into the space around them, seeking for supports to climb up. The illness was deeply frustrating, but the project also appealed to Darwin: "it is just the sort of niggling work that suits me," he wrote. He asked Hooker for more exotic species to observe. His ailments had hamstrung his ability to live as he usually did: busily setting up experiments and keeping up with his many correspondents.[1] He was only fifty-three, but he had been forced to live slowly—to become more plant-like—as he healed and recovered his strength. This stillness had opened his already attentive mind to watching his plants with unprecedented patience. It had allowed him to see them more on their terms, to experience plant life at plant pace.

Of course, being the avid naturalist that he was, Darwin did not allow himself to become inert. He spent four intense months captivated by the tendrils of the cucumber plants. When he became mobile enough, he sat out in fields and watched hop shoots make their ascents up growing poles. He brought them inside to join the potted cucumber and clematis creepers making a latticework of stems across his windowsill, ambitiously seeking the light outside. He started to attach small weights to them to slow down their movements and daub markings on them to track their progress over time. By the end of the summer, he had written a sizeable paper, published as a 118-page monograph by the Linnaean Society: *The Movements and Habits of Climbing Plants*.[2] In it, Darwin pointed out evolutionary connections between the way the cucumbers climbed using tendrils like springs that coiled around objects, and the method used by the clematis, which was to clasp on tightly with "hooks." They were two ways of solving an important evolutionary problem: how to reach light without a rigid stem.[3]

Figure 1

Darwin's amusement with his "tendrils" reveals the kind of mental shift needed to enter into the plant world, to put oneself imaginatively into the being of an entirely different kind of organism. He was referring to the tendrils of his plants, of course, but over the weeks of his confinement, he came far closer to those plants on their own terms than the taxonomists concerned with nomenclature, or the plant physiologists dissecting and observing the minutiae of plant bodies in the lab. Knowing the names and family trees of plants in fine detail, or the physical mechanisms by which plants operated, did not reveal much about them beyond the material. These were ways of *looking at* plants. Darwin wanted to *see* them. And what he saw certainly wasn't boring. Some of his plants genuinely surprised him. They could do things that were not simple and not always slow. Sometimes, what they did was shockingly fast.[4]

Darwin went on to develop a way to record what he saw with the naked eye. He would place a plant in between a sheet of paper and a glass plate. He marked a reference point on the paper and attached a filament to whichever plant organ he was interested in. At regular intervals, he would line up the end of the filament with the fixed reference point by eye and mark the position on the glass plate. By connecting the dots on the plate in order, he drew the movement of the plant organ, which became far more visible to the naked eye because it was now magnified many times. This was a remarkably creative way of capturing plant movement that could be taken in by the human eye before the days of time-lapse photography. Darwin could even "zoom in" on movements by varying the distance between the plate and the plant. By moving it further away, he increased the angle at which the dots all lined up, thus making small movements appear as larger distances in his tracings. By watching their movements and growth, Darwin began to pioneer an understanding of plant "habits." He understood before anyone else the changes that plants make in their physical positions and shapes as *behaviour*, in the same way that the movement of animals would be. Growth is slow in all organisms, but for plants, almost all of their movement occurs as a result of their patterns of growth and development. This was something that the approach of the experimental plant physiologists in the lab erased in their purely mechanistic approaches.[5]

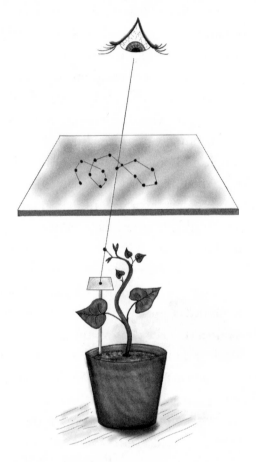

Figure 2: A cartoon of Darwin's glass plate method for observing plant movements. Darwin visually aligned the bead of wax on the plant with the dot on the card behind and marked its apparent position on the glass plate. Over time, he joined the dots in order and built up a tracing of the circumnutating movements of the plant.

To understand plant intelligence, we must observe plant behaviour as carefully as Darwin did himself. We need to see more than just the fast, snap-shut movement of the Venus flytrap or folding leaves of the mimosa with which our naked eye can engage. Virtually no growing part of any single plant remains still. All plant organs move: from root tips and tendrils to leaves and flowers. They all sway in circles as they grow, a pattern which Darwin called "circumnutation" (from the Latin *circum*—round, and *nutare*—to nod). With his glass-plate technique, he was able to trace hundreds of movements in the stems, flower stalks, leaves and leaflets of his plants, crystallising their exploratory perambulations in staccato lines.

Before I began investigating plant intelligence at MINT Lab, I moved to Edinburgh for a year to work in three different departments—in philosophy, psychology and plant biology.* It was the confluence of these three disciplines that would eventually lead me deep into my study of plant cognition. It took this combination to allow my ideas to blossom: too narrow a focus and I would never have made the connections that I did. Just after I had arrived, I had an experience—a flash of insight—which precipitated my journey down this path. The flat I had moved into could not have been better situated. It had a bay window which looked out both on to Edinburgh's iconic Arthur's Seat and Hutton's Section, the site where James Hutton made his startling breakthrough discoveries on the dynamic nature of rock formations in the

* Funded by the Spanish Ministry (Estancias de Movilidad de Profesores e Investigadores Senior en Centros Extranjeros-Educación, Cultura y Deporte), I spent a one-year sabbatical leave at the School of Philosophy, Psychology and Language Sciences, University of Edinburgh, under a triple appointment (PMARC—Perception Movement Action Research Consortium with professor David N. Lee; EIDYN—The Edinburgh Centre for Epistemology, Mind and Normativity with professor Andy Clark; and The Institute of Molecular Plant Sciences with emeritus professor Tony Trewavas.

mid eighteenth century.[6] It was also a stone's throw from where Darwin himself had lived as a medical student in his late teens. Though he was there to study the human body, Darwin could not help taking a holistic view of life, peering into the interconnected worlds of entirely different kinds of organisms. He would often attend botany lectures by John S. Henslow and go on plant-hunting expeditions.* He could frequently be found walking along the Firth of Forth with his mentor Robert E. Grant to look at sponges. At the time, these creatures were thoroughly mysterious, thought to exist on the boundary between animals and plants. Darwin would come to picture life not as a hierarchy, but as a branching, interconnected tree.[7] He could see that even apparently inert organisms were worthy of attention, all engaged in a brutal struggle for life.

Despite its exceptional location, my flat lacked furnishing. Rather than decorating properly, I bought a record player to make it feel like home, along with one record: Ella Fitzgerald singing *The Cole Porter Songbook*. One day I sat at the bay window looking out at Arthur's Seat and watching the slow dance of the large potted telegraph plant I had sitting by the window. I listened to Ella singing in her rich, velvety tones: "birds do it, bees do it, even educated fleas do it," playing the song again and again. The different animals and plants ran through my mind as she named them, from the Boston beans, sponges and oysters to the clams, jellyfish, electric eels, soles and even goldfish cloistered in their bowls. Ella was singing about their sexual entanglements, but in my mind, what all these organisms were "doing" was something far more impor-tant than romance and sex. They were not just "falling in love," they were all exhibiting their own kinds of intelligence.

---

* Henslow would later turn down the invitation to sail on the *Beagle* in Darwin's favour. We owe him quite a debt of gratitude.

My mind flitted between snapshots of their startling abilities: the communication between ants or termites, the forethinking abilities of fruit flies, the ingenuity of worms dragging different shapes of leaves into their burrows. Just as the enticing enmeshments of sexual reproduction were not limited to humans, so different elements of cognition were also found throughout all lifeforms. Each of these organisms had their own kind of intelligence, including the non-animal ones.

As I listened, I let my mind go through a strange transformation. It was as if I clambered through the limbs of the tree of life, passing each of our animal forebears—the primates, early mammals, bony fish, all of the invertebrates that Ella sang about—as I went, back in evolutionary time to the ancient single-celled common ancestor of animals and plants nearly 1.5 billion years ago. I then began a mysterious ascent through an alien photosynthetic dynasty, all the way to the family of the plant I was watching.* My image of myself changed as I took this journey. My animal frame of muscles and skeleton, controlled by a cranium-bound brain, dissolved into a slow, flexible, elongated being with an entirely different kind of awareness of the world. In my imagination, I became plant-like, but I know that this was only a game of make-believe. It was a mental experiment that might give me clues to understanding what I was seeing as I watched the plant in front of me: the achingly slow, circling dance of bending stems and tilting leaves in their balletic efforts to catch each drop of sunlight.

A quote surfaced in my mind as it sat quietly in this vegetal

---

* Animals and plants originated from a single-celled common ancestor 1.5 billion years ago. This was probably motile. The lineage that led to plants engulfed a smaller photosynthetic cell which eventually developed into the chloroplast. Moving to acquire energy was no longer necessary. See McFadden, G. I. (2014), "Origin and evolution of plastids and photosynthesis in eukaryotes," *Cold Spring Harbor Perspectives in Biology* 6: a016105.

state, from *The Secret Garden* by Frances Hodgson Burnett, which I was in the middle of reading:

> I don't know its name so I call it Magic ... Everything is made out of Magic, leaves and trees, flowers and birds, badgers and foxes and squirrels and people.[8]

This "Magic," as Hodgson Burnett would put it, is something common to all living things. It is the very stuff that animates all life. Leaves, trees; flowers, birds; badgers and foxes and squirrels and people all exist along the same continuum, animated by the same essential things, expressed in the particular ways that their evolutionary journeys have elicited. They exist less on a hierarchical "tree" of life, more in an "adaptive landscape" in which each species is incrementally climbing its own evolutionary slope. Given this, can intelligence have originated *only* in animals? I would argue not.

The idea of a landscape was described by Richard Dawkins in *Climbing Mount Improbable.*[9] He uses it to convey how the apparently impossible development of complex adaptations and unimaginable diversity has happened through tiny steps over vast swathes of evolutionary time, each species winding gradually towards an evolutionary summit. As Dawkins puts it, this "slow, cumulative, one-step-at-a-time, non-random survival of random variants" was what "Darwin called natural selection." Each species has its own slope, there is no one ultimate peak. These mountains cannot be climbed suddenly. You cannot jump up a hill, and neither can you reach another peak by going downhill. Once you begin an ascent, there is no going back: "species can't get worse as a prelude to getting better." There are many peaks, many ways of solving the same problem or being highly adapted to the environment. To take a classic example, eyes of different kinds have evolved over forty times. Each type of eye is a slightly different solution to the same problem: how to turn light into information about an organism's surroundings.[10] This metaphor might be more helpful than the image of a tree in helping us to overcome our perceptions of "higher" and "lower" forms of life. The tree depicts branching relationships over time, but it is misleading in combination with our inherent tendency to ascribe values to things. The idea of a mountainous landscape, paradoxically, creates a level playing field, each species faced with its own task, beginning from the same substrate and climbing busily away.

Playing with time

The distinction between *looking at* and *seeing* other organisms, especially those so very different from us as plants, is more difficult to make than it might sound. We can all gaze at a plant, we can create detailed taxonomies of all the species or uncover the physiological mechanisms by which it grows and develops. But *seeing* what a plant is really doing, *understanding* it, is much more subtle. It requires a change of perspective. It doesn't happen naturally for most of us when it comes to plants, for all the reasons we saw in the last chapter. And what we might manage to see is only the very surface of something that may be highly complex. If we want to understand what is really going on when plants live their lives—to get to grips with plant intelligence—we cannot observe it directly. We can only observe how a plant grows and develops from seed to adult, and then *infer* the nature of the intelligence that drives it. But this cannot be an act of fanciful storytelling.

Darwin himself was ridiculed by the establishment for what they saw as his lack of rigid experimentation, despite the fact that his ideas were based on careful, objective measurements.[11]

Obstacles arise at every level: in order to have insight into plant intelligence, we must make observations of plant behaviour in such a way as to tease out what is going on underneath. In order to see plant behaviour, we have to perform some kind of manipulation of plant growth to reveal to our own animal senses what is going on. This will unavoidably affect the picture in some way, even making the plant seem more "animal like." If we don't proceed very carefully indeed, we risk simply speculating or, perhaps worse, reducing the complexity of plant behaviour to nothing but purely physiological reactions. Seeking a plant's perspective is a tricky business.

Behaviour is the golden thread that can lead us to the centre of the complex maze of plant intelligence. We need to find ways of bridging between the perception of our animal senses and the ways in which plants behave. The tools we may use to do this very much depend on what we are looking at. It is most often done by transforming the movement-by-growth that makes up much of plants' overt behaviour into something easy to take in. The timescale of our perception operates on an image length of about one-tenth of a second (we are able to process some ten to twelve images per second on average), far shorter than the timeframes according to which plants grow.[12] One solution is to compress time, an idea which has long entranced me. Before I left Spain to travel to Edinburgh, I became fascinated by pinhole cameras, in which a simple photographic paper is exposed over many minutes through a tiny hole. The first photograph I took was of my sister lying on a bench in front of a sunlit bay. My sister stayed perfectly still over the minutes I exposed the photographic paper, but her skirt was blown by the wind, and the sea was whipped into small waves. As a result, she was crisp and defined in the resulting image, but her skirt and the surface of the water were blurred. The image had captured the different scales of movement over the time.

Figure 3

I set up a dark room in my Edinburgh flat, and took pinhole photographs of different plants. In the cool light of northern Europe, I needed to expose the paper for much longer than in the Mediterranean sun. I would sit perfectly still for a full fifteen minutes while the paper was exposed, capturing the movement of the plant in that time. What I was left with was an image that telescoped the time, as if all the frames from a film had been amalgamated into a single image. With a pinhole camera, different timescales can be collected in one image. Because there is no lens, the whole visual field is in focus, as compared with a more complex camera where only one slice of the image a particular distance away is in focus. With a pinhole camera, everything that happens imprints onto the photograph. Objects that stay still over the exposure time will be sharp, those that move will be blurred, their movements over time aggregated into a flurry in the image. It is an extension of the naked eye's experience turned into a photograph.[13]

I began to use new tools, of a kind to which Darwin had no access, to push beyond the limits of my perception. I began taking time-lapse photograph sequences of plants growing in my garden. One exotic climber, *Araujia sericifera*, grows especially well there and is becoming invasive across Europe. It is called the "cruel vine" because it traps pollinators such as moths and butterflies by their long tongues in its rigid flowers. I did nothing to the vines' environment, added no poles or complex experimental set-ups—I simply took time-lapse footage of them as they pirouetted around looking for supports or tangled themselves across my olive and orange trees. Eventually I took this time-lapse project into the lab, and it became the work of MINT Lab. I realised something through the transition which was hard to admit, even to myself: no matter how much focus and imagination I could muster, I was aware that when I watched plants with my naked eye, I was less immediately impressed by their abilities than when I watched back time-lapses that I had taken. The plants at artificial speed still appealed to my biased animal mind in a way that effortful, slow observation could not.

To appeal to our senses, and make behaviour more intelligible, we can take snapshots of plants over many hours and telescope them into a brief clip with time-lapse photography. If a climbing plant is circumnutating around its rooted point looking for a support, the naked eye will find it hard to focus on the movements over hours, but will appreciate a sequence which shows the stem circling over a few minutes. This technology was in existence not long after Darwin's observations with his glass plates. Between 1898 and 1900, just after the Lumière brothers invented *le cinématographe*, the German botanist Wilhem Pfeffer had already made pioneering studies of plant movement by assembling a "time-lapse clip."[14] He made mesmerising clips of tulips flowering with shuddering speed, folding and unfolding mimosa leaves, the spreading roots of germinating seedlings, rotated plant shoots growing up against gravity. The ability to visualise plant behaviour was already around long before we started to pay any attention to it.

The results of this kind of technology have become the key to turning plants into the protagonists of nature documentaries, in a way they never would have been in the past. Many of Sir David Attenborough's programmes have used time-lapse sequences over hours, days and even whole seasons to show the private lives of plants in a way that is just as exciting as the footage of animals we are more familiar with. The germination of seedlings on the rainforest floor over a few days becomes a shivering, agitated race for life spanning a few seconds; the shift in colours and foliage over the seasons becomes a rippling psychedelic display. What would take effort to notice with the naked eye, or even be practically impossible to see, becomes riveting and immediately gratifying. In the MINT Lab, we distil this effect for scientific study by focusing on individuals. We place the plant at the centre of a cylinder with a camera fixed above it, taking one picture per minute. Played at twenty-four frames per second, we can see twenty-four minutes of activity in just one second of footage, so many hours can be compressed into a few minutes. We can even use infrared light to keep recording in darkness and watch plants around the clock.

Plants, however, don't always need to be sped up. To see what they are doing sometimes requires us to slow down their movement. The shutting of the Venus flytrap's leaves or the rearrangement of stamens in response to the trembling touch of pollinating insects might happen far more quickly than our eyes can take in. For these cases, we need high-speed film cameras which can capture around one or two thousand frames per second. *Catasetum* orchids, for example, catapult a sticky pollen organ called a "pollinium" onto unwitting insects visiting the flowers. This happens so quickly that the insect has no time to get away before it is saddled with the flower's cargo and has to fly off heavily laden, ideally to another flower where the pollen will find fertile reception. This catapulting happens at an astounding

3 metres per second (10.8 kilometres per hour). For an organism without nerves or muscles, this is incredibly fast. Blink and you'll miss it. Even if you don't blink, you will not be able to actually see the pollinium's brief arc of flight. The top speed ever recorded for a pollinium was 303 centimetres per second, making it one of the fastest movements in the entire plant world.[15] It is no wonder that Darwin called them "the most remarkable of all Orchids."[16]

The complexity of the technology we use to allow these shifts in perception does not mean that it will automatically make the right kinds of measurements of intelligent behaviour. Plenty of extremely high-tech observations can be made of things which are entirely inanimate. Take the Hubble Space Telescope, launched into Earth's orbital field in the 1990s. An incredible feat of engineering, it can take high-resolution images across a broad spectrum of radiation, from the ultraviolet to the near-infrared, allowing us to see deep into space. It is one of the great achievements of human ingenuity, but it records inanimate matter. To capture *behaviour* with technology in a way that reveals the intelligence behind it, we need to be smart with how this technology is deployed. And that is why we need to think very hard about how we approach our explorations. We need to use the right imaging tools to tune in to what needs capturing and understand the effect that the recording technology might have on the plant and the limitations of these methods. Carefully tailor-made experimental designs are needed to fashion results that will reveal what is going on underneath.

Any kind of technological mediation between us and plants is going to come at a cost. Sampling, recording and editing are all going to introduce bias into our experiences in one way or another. Time-lapse can turn the apparently inanimate into movement in our eyes, growth into behaviour. It brings what is barely perceptible comfortably within reach of our sensory systems. But it has to be deployed with caution and care, so that the results are not a grainy, incomplete picture of what a plant is doing, like an overly filtered photograph on social media. It would be easy for us to assume that taking one frame per minute might be enough to give us a suitably high density of observations to trace all the movements of a plant seeking a climbing support over several hours. But we would be wrong. Taking one photograph every minute means that you miss 59/60ths of what the plant is doing, and not all of the plant's movements are slow.

Take common beans, for example. They might spend about an hour doing a ponderous rotation around their surroundings, all of which could be sped up into a few seconds of footage (sixty frames captured in an hour, played at twenty-four frames per second). But they can also be devilishly quick. Occasionally I have come across particular bean plants that have taken me completely by surprise. Plants are individuals after all, they do not all behave the same way. One bean that I liked to call "Usain Bolt" was able to make a rapid "grab" for a pole and lasso it tight without even having made contact with it. This "grab" was not part of the usual anticlockwise rotation and it was very fast, lasting only a few minutes. So fast, in fact, that it did not show up on the time-lapse footage in any detail. Quite aside from knowing *how* the bean made this grab, we could hardly see *what* it was doing because the time-lapse misses so much. Like a poltergeist in a horror movie toying with protagonists who set up secret cameras to see who is moving all the furniture, it is almost as if the beans erase the footage from the time-lapse sequence. They achieve something unexpected and mysterious right under our noses, despite the cameras. In reality, though, we allow ourselves to be lulled into the false reality that the continuous stream of frames represents continuous time. In fact, the brief action shot was never captured by our equipment. In this case, the naked eye could see the grab over a few minutes much better than a time-lapse sequence could.

Making plants into animals

Several years ago, in the city of Hyderabad, the capital of southern India's Telangana state, a strange palm tree enjoyed a brief period of local celebrity. From the morning, the tree would start to lean over, like a drunk propping up a bar. It would continue a slow-motion topple to an ever more precipitous angle through the day. By the evening, its uppermost leaves would almost be touching the ground. Impossibly though, after dark, the palm would begin to right itself. By midnight it was again perpendicular, standing upright at its full ten-foot height as if nothing had happened. The local people took this mysterious daily feat as a sign that supernatural forces were at work. They flocked to the tree to pray, thinking it a conduit of some divine communication.

A professor at Osmania University in South India wrote to me, along with a number of other academics, seeking advice on how to communicate a scientific explanation of the tree's unusual gymnastics to the locals. He was concerned that the leaning palm might become the focus of some kind of cult. One of his correspondents suggested an explanation, based on plant physiology and the unusual situation of this particular palm. It was possible that during the day, the tree lost water evaporating in the heat of the sun, making it lose its turgidity so the trunk became flexible and the crown drooped over. During the night, the tree soaked up water from the well next to it, becoming waterlogged, which allowed it to regain its upright posture. The trunk might also have been damaged by parasites or by the constant hinging, making it even more flexible. This explanation was not simple, and was certainly not obvious to the local people. To them, movement was something that plants did not do: this tree's motion must mean some kind of supra-natural animism was at play.

The material cause of the palm's bending was of only mild interest to me; people's inclination to interpret the palm's activities as something beyond the natural was much more intriguing. Other palms have been seen doing similar things in the past, and observers have come up with a variety of explanations. The early twentieth-century Indian polymath Sir Jagadish Chandra Bose recorded the daily bending movements of the "Praying Palm of Faridpur," a date palm in Bengal, and attributed the movements to complex interactions between responses to gravity and temperature, involving oscillating electrical signals in the plant. Bose eventually developed a hypothesis about how plants explore and respond to their environments which became one of the pioneering moments of early plant physiology.

As we have seen, to the human mind, movement and intelligence are inextricably linked. Which means that plants, seemingly motionless, are difficult for us to consider as beings with intelligence. Ironically, though, we are extraordinarily good at projecting all sorts of grand plans onto random moving objects. This tendency is not dissimilar to the way that we can see faces in any collection of lines and shapes that could conceivably resemble two eyes and a mouth. The extent of our tendency to focus on apparent "behaviour" was demonstrated by experimental psychologists Fritz Heider and Marianne Simmel in a study in 1944.[17] They showed thirty-four undergraduates a black and white video of two-dimensional shapes—triangles and circles—moving around for a minute and a half. The psychologists then asked the subjects to "write down what happened." Most of the accounts read like soap opera synopses. The majority described the shapes as if they were either men or women, imbuing them with goals, plans, and the ability to react to the actions of those around them. The shapes became characters in stories.[18]

Heider and Simmel carried out the same experiment with thirty-seven undergraduates and asked them to describe the personalities of the shapes. This time, the shapes were not only animate, they had relationships and emotions. They were "heroic," "cowardly" or "mean." When two circles spun around one another, it was an expression of joy. When a circle lurked inside a rectangular outline, it was fearful of the aggressive triangle prowling the perimeter. Add movement, and monochrome shapes can become like people in the human imagination. This imaginative capacity is valuable. It is what makes us the social beings we are, able to hypothesise about the mental worlds of others and interpret what they do in meaningful ways. But it is also a source of misleading mirages when it comes to understanding beings that exist in the world in a way that is alien to us. Plants are animate but they are not animals. We can't look at their faces to understand what is going on internally. We need to make efforts to see plants and their subjective experiences on their own terms.

There is a real danger here that could mar our efforts before we have hardly begun. If we want to understand plants, we need to both avoid anthropomorphising them and being too zoocentric in our approach. As we saw with the Hyderabad palm, we are likely to interpret as being beyond a plant anything that seems too active for it. As we saw from the shapes experiment, we can anthropomorphise almost anything. It is in our nature to draw analogies between things we see in ourselves and things we see in the world around us, to extrapolate from the familiar to the unfamiliar, the close to the distant. Our tendency to project our own experiences onto other organisms, to give spirit to the inorganic world, has given rise to the multitude of rich mythologies and animistic religions throughout history. We unavoidably use ourselves, our internal subjective experience, as the first reference point for understanding the world.[19] But we cannot learn about the world by analogy: these kinds of assumptions are not based on data. They are simply projections of ourselves onto the outside world. At best, they are very difficult to prove to be correct. At worst, they are very likely to be wrong and misleading, especially if the distance they attempt to bridge is too far.

There are two extremes: the anthropomorphic tendency to see ourselves in things that are entirely unrelated, and the anthropocentric refusal to acknowledge continuities that exist between ourselves and other forms of life. A loving pet owner might imagine their snake is "happy" when it gets fed, because the owner's own mood is lifted by food. Whether or not the snake has this kind of emotional range, we don't really know. Reading a snake's expressions is much trickier than reading those of a domestic dog. By contrast, we might be so afraid of anthropomorphisation that we withhold acknowledging the emotions of other organisms at all. In *The Expression of Emotions in Man and Animals* (1872), Darwin included a plate of a cat rubbing against someone's leg, captioned "a cat in an affectionate frame of mind." Statements such as this were criticised by psychologists as being anthropomorphic well into the twentieth century. But Darwin would have responded that assuming only humans had emotions was anthropocentric. He argued that such "affectionate" behaviour must be linked to the internal state of the cat, intended to have some kind of effect on the organisms around it. It followed that mental states, and their expression, were integral to the interactions between social animals.[20] The ghost of anthropomorphisation and accusations of pseudoscience hampered the explorations of the emotional capacities of other animals for a long time, but the tide is turning.[21]

This double-edged sword is even keener in work on plant intelligence. Plants may be harder to immediately anthropomorphise, but they are also much stranger and further away from us. It is a much greater challenge to both observe and remain objective. Our work at MINT Lab has certainly had its critics, as we shall see in Part II. In the *New Yorker* in 2013, the brilliant writer Michael Pollan quoted plant physiologist Lincoln Taiz, who had asserted that calling climbing beans "intelligent" ran the risk of "over-interpretation of data, teleology, anthropomorphizing, philosophizing, and wild speculations." Taiz suggested that we at MINT Lab were falling prey to "animism." Among many other questions, he asked what sensory modality allows the bean to perceive the pole (if it does at all). How does the plant control the sweeping approach? Isn't the "intelligent" aspect of this all in the eye of the beholder, doubtless prone to anthropomorphisation?

I couldn't agree more. We must be very careful and avoid overexcitement when investigating plants' cleverness. This is science, after all. Excitement is part of what motivates us, whatever our scientific pursuit may be—we want to *know*. But caution is also what makes science such a solid foundation for knowledge. When it comes to plant intelligence, I happily assume the burden of proof. It takes some mental discipline to prevent oneself from over-interpreting the navigation skills of the bean plant, dodder or *Passiflora* tendrils as goal-directed. But we also must not be zoocentric. Taiz and his team suggested that "time-lapse videos of growing roots or twining stems, which have been sped up to make them look more animal-like, do not constitute evidence for consciousness or intentionality."[22] Their interpretation epitomises a widespread misunderstanding: that intelligence and consciousness are inextricable from the type of responses that can be detected by our own scale of perception, i.e., rapid movement, such as that found in animals. They are not. It is not the speed of behaviour that constitutes evidence for intelligence. We are not artificially trying to make plants seem like animals by using time-lapse, we are simply making plant behaviour easier to perceive as a result of collapsing time, rendering that behaviour visible to us so that we can uncover the intelligence underlying it. Time-lapse reveals complex patterns and flexibilities in plant behaviour that we would not otherwise be able to see, just as slowing down the rapid movements of some animals, such as the flight of birds, allows us to see them properly and understand them. I would respectfully suggest that, paradoxically, it is their own zoocentric viewpoint that is fueling their accusations of zoocenterism on our part.[23]

What does this mean in practice? When we look at a time-lapse sequence of a vine exploring its environment, we can see that it behaves in a systematic way. It can reach for a surface, test if it is suitable and, if not, can withdraw; it can adjust its position precisely, and repeat the cycle if it needs to. Our instinct is to interpret this process as if the plant has intentions, that it has a plan for what it is doing. This is, of course, an anthropomorphic perspective. However, our intuition about what is happening is our natural response to what we are seeing from a plant that has to navigate a complex landscape of environmental threats and opportunities, that has not only to be responsive and flexible in its behaviour but also proactive and predictive. We can investigate this in different ways: time-lapse photography allows us to observe plant behaviour; investigating the physiology of the plant reveals parts of how plants function on a biochemical and developmental level. But one would still miss the underlying plot.[24] Understanding the apparent meaningfulness and intelligence that the plant's behaviour reveals requires a different approach, one combining careful plant science, cognitive science and philosophy. And this is where our work at MINT Lab comes in.[25]

If we are careful enough with the way we make observations we can begin to tease out evidence for intelligence from observations of plant behaviour. We want to borrow paradigms from animal cognition studies when designing experiments in plants, to provide a theoretical framework for guiding our investigations. We will not understand plant intelligence if we limit ourselves to cutting plants up and working out the details of their physiology, nor will we be able to infer what is going on by just looking at their behaviour. It would be madness to disregard the decades of work that have gone into developing ways of applying experimental set-ups to animals which reveal the intelligent underpinnings of animal behaviour. Not only that, but there are valuable and valid analogies to be drawn between plant and animal cognition. But that does not mean we want to make plants into animals.

Seeing not looking

I began this chapter with a brilliant but bed-ridden naturalist watching the plants growing on the windowsill of his room, and arrived at high-tech cameras making seamless time-lapse sequences in the lab. This is how the technology has progressed, but to see plant intelligence we need to use all the ways of looking that we have. Apart from clever science, if we wish to learn to *see* plants rather than simply *look* at them, we need to retain something of the personal connection that Darwin developed with his plants in the nineteenth century. Only the naked eye can teach us about *this* plant, in *this* place at *this* time. Technological tools are invaluable aids, but they have limitations and must be used wisely. We have to ignore the focus on taxonomy and botanical nomenclature that has dominated the study of plants for so long—labels and abstract groups are not going to tell us much about the intelligence of individual plants.

Instead, we need to connect to the concrete and specific nature of particular plants and their worlds. We must acknowledge that plants are animate without slipping into animism; relinquish the anthropocentric hegemony on intelligence by not being anthropomorphic. This discipline, of applying cognitive studies to a whole new realm of organisms without bias, cannot but make us better at looking at our *own* cognition with an objective eye and with new perspective. If we are successful, we might manage to uncover what is really going on behind the spectacular feats that plants can achieve.

CHAPTER THREE

SMART PLANT BEHAVIOUR

Root-Brainstorming

Long before I became ensconced in my research at MINT Lab, I found myself in the village of San José in southern Spain, watching an unusual ad-hoc piece of cinema. I was with two plant scientist friends of mine, František Baluška and Stefano Mancuso, who were part of a small group I had invited there for the weekend to brainstorm together about plant "brains." We called the gathering, somewhat cryptically, the "San José Root-Brainstorming Meeting," playing on their preferred hypothesis as to where the plant "brain" might reside. Just like Darwin, František fervently believed that plant processing occurred at the root tip. He had spent many years researching this idea at the Institute of Cellular and Molecular Botany at the University of Bonn. We wanted to see where our combined interests and expertise could take us in pushing the ideas further.

Our discussions took us into the early evening, so, as dusk approached, we headed to a seaside bar to refuel and have a drink. Stefano had recently time-lapsed a rather vigorous variety of pole bean, *Supermarconi*, at the International Laboratory of Plant Neurobiology in Florence.[1] He carried the precious footage with him on a USB stick. Luckily, I always carry a small portable projector with me, so we decided to set up a makeshift cinema right in the bar. The barman helpfully cleared some wall space for us, and the phantom bean plant began to snake up the wall, luminous between the shadowy shelves of glasses and spirits, appearing to reach effortfully for a pole. We watched the time-lapse numerous times over, totally enraptured, much to the confusion of the other customers. They could not quite seem to see what it was about this footage that excited these three clearly very eccentric men at the bar. But we, at least, were absolutely thrilled. It was clear that there was far more to the bean's activities than met the eye.

The next day, we found ourselves on the beach, tracing experimental designs on the wet sand with a cane, letting our ideas spill out onto the particulate, borderless canvas. We did not want to get ahead of ourselves, but as we scoured the sand with sketches of what came to mind as we talked, a whole host of new ways of looking at the climbing of the bean unfurled before us. The beach temporarily became an expanse of tendrils and stalks, arrows and lines hypothesising their actions, all of which would be washed away at the next tide. František and Stefano knew all about how the beans "moved," and yet something about it eluded me.[2] The "motor zone" between the tendril at the end of the plant and the vertical stem was where the control was, allowing them to circle around—the circumnutation that Darwin described—and to finely control the nature of this movement.[3] Circumnutation is not an automatic pattern; the plant can adjust what the tendril does. Cells in this motor zone act like hydraulic pumps, expanding and contracting on either side of the stem. Charged particles are moved around in wave-like patterns between cells, and are followed by water which changes how turgid the cells become.

This effectively increases or decreases the relative lengths of each side of the stem, which causes the tendril to move.[4] Think of it a little like a Mexican wave, a smooth, rhythmic pulse of liquid inflation and deflation of the cells. This had to be under the plant's control, but I could only begin to imagine what was behind it at that point.

Darwin described these movements, in *The Movements and Habits of Climbing Plants*, from his studies of the sophisticated movement of *Ceropegia*, a common ornamental plant, climbing a support. He likened the plant's movement to a rope being swung in ever-shifting arcs until "it again came into contact with the stick, again slid up it, and again bounded from it and fell over to the opposite side."[5] To us, the bean appeared as a rope-throwing cowboy lassoing a pole or a fly fisherman casting his line forward and back, getting closer and closer with every cast. After the first overhead cast, the bean keeps casting until it reaches the support. It pauses slightly at the end of the back-casting movement, and swiftly begins the forward approach once more. All, apparently, with the aim of locating the pole and making a final, targeted "grab" for it.

Stefano's time-lapse film and the excited discussions between the three of us over that weekend stayed with me for years. They inspired my initial work at MINT Lab, where we began to explore these ideas experimentally. We first considered analysing the movements of *Ceropegia*. I liked the symmetry of studying the same plants as Darwin, thinking that he would have encountered them during his voyage on the *Beagle*. We know that he passed Tenerife, and some of the hundreds of known species of this plant are distributed widely in the Canary Islands, so it would be simple to bring their seedlings to the lab. To my dismay, though, I discovered that he never even set foot in the port of Santa Cruz. Due to an outbreak of cholera, the crew was quarantined for almost two weeks, and the captain decided to move off without touching land.[6] Eventually, we decided that the common bean (*Phaseolus vulgaris*) would be a good candidate for the next stage of our research, not least because its circumnutation is comparatively simple, while that of *Ceropegia* is much more complex.

We started studying seedlings of *Phaseolus* beans through controlled time-lapse photography in the lab so that we could see their growth, which was, of course, also their *behaviour*. In 2016 we made a custom-built booth for growing and time-lapsing the beans as they appeared to search for the pole, which was placed 50 centimetres away from them.[7] Graphing the footage we took, we could see it all at once, many hours of movement condensed into a single image. And it revealed a striking pattern.[8] The bean revolved in the space around it in increasingly larger arcs, until it reached the pole.

To cut a long story short, the tip of the shoot moved roughly in a helix, its movement shifting gradually from circular to elliptical as the shoot grew. The bean performed twenty-one cycles, taking an average of 117 minutes per lap. The first cycle of revolution was the shortest, lasting 98 minutes; the longest 154 minutes. This whole pattern of movement is far more cumbersome (and interesting) than it might first appear. At one point, the shoot skipped the second half of its helical trajectory, and cut across the middle to fish for the pole directly. At last, we were able to actually *see* plant behaviour.

From adaptation to cognition

At MINT Lab, we were interested in understanding how the plant reached for a support—what made it *goal-directed*. Such target-oriented behaviour appears to require a fine-tuned cellular machine able to orchestrate it all. And yet the question remains: what is the nature of this machine? Is it merely automatic or does it involve complex processing akin to that we know in animals? To prevent us from wilfully blinding ourselves with the biases of our history, it is vital that we engage in this research with open minds. We do not yet know the terrain well enough to be able to *expect* what we might find.

Critics of our work at MINT Lab suggest that these kinds of observations show nothing more than sophisticated adaptations, behaviours that boil down to automatic responses to stimuli. The flowers of orchids, with their astonishing array of different colours and shapes, are exquisitely formed to deceive insects and load them with pollen and yet are examples of natural selection, not cognition. Critics argue that what we see in climbing beans and our other experimental subjects does not signify cognition-driven behaviour in any meaningful sense, and cannot be interpreted as such. We would argue that they are wrong, that this behaviour cannot be explained fully as mere reflex responses. There must be more to it. But at the same time, it is up to us to prove it. So let's unpack this complex issue gradually. I will see if I can convince you of my argument that the beans are doing more than we imagine. That the climbers are not only well-adapted organisms, but that their abilities, and those of other plants, are due *to plant cognition.*

To distinguish between these positions, we need first to understand what the subtle difference is between adaptation and cognition, or rather, what cognition is that cannot be explained only by adaptation.[9] Because, of course, cognition is also adaptive, allowing plants to better inhabit their environments.[10] "Adaptation" often refers to something that is an automatic response to a particular input. It is *genetically encoded,* by virtue of that characteristic having been largely advantageous over evolutionary time. It is *reactive*: the stimulus has to happen to trigger the response, and it is always roughly the same response. Like the motion detector in a garage door which controls whether it continues to close or not, behavioural adaptations respond to particular conditions in the environment in particular ways. There is not much flexibility, only the mechanism that has been wired in genetically. This means that it takes very little processing power. Think of the knee-jerk reflex that happens when your knee is tapped: your leg kicks out even before the signal has reached your brain as a result of a closed circuit of neurones. No computation is needed, so it is very rapid, preventing you from falling over if you bump into something. But this means you also can't really control or modify the movement.

Cognitive behaviour *is* adaptive, but it is much more besides. It is *anticipatory*, allowing an organism to optimise for *future* changes in the environment. It is *flexible*, responding to multiple different factors and with multiple different manifestations. It is also *goal-directed*, aimed at making a change in the environment or in the organism's state, rather than simply responding to it. These qualities require much more than "knee-jerk" reactions. They need to use information from many sources and in different parts of the plant, from root to shoot, all of which must be *integrated* to allow a coordinated response.[11] They can be improved by learning over the lifetime of the plant, to better shape what it does for the future. Some of the ways that plants have of enacting such cognition-driven behaviour are through growth, some through rapid movements, and some through the release of powerful chemicals that affect organisms around them.

In *The Power of Movement in Plants*, Charles Darwin and his son Francis wrote that "there is no structure in plants more wonderful, as far as its functions are concerned, than the tip of the radicle." The "radicle" is the growing tip of a germinating root, which encounters different aspects of the outside world—from light and gravity to physical obstacles—and makes choices about how to engage with them in order to successfully find its way underground. The Darwins summed up: "Two, or perhaps more, of the exciting causes often act simultaneously on the tip, and one conquers the other, no doubt in accordance with its importance for the life of the plant." This war of causes, and the resulting behaviour, is where plant cognition can be deciphered.[12]

Walking palms and cannibal caterpillars

To give us a clear picture of what might distinguish these kinds of behaviour in plants, I want to move gradually from examples of the simply adaptive to those of the potentially cognitive. Astounding cases exist all along the scale of underlying complexity: we have to look closely to know what is behind them.

Even physical adaptations can be incredibly striking. In the humid tropical rainforests of South America there is a plant called the walking palm, or *Socratea exorrhiza*.[13] It is a waif-like thing, with a trunk averaging only 12 centimetres in diameter, but grows all the way up to about 15–25 metres in height; it seems impossible that it does not simply fall over. It has very unusual roots though, a basket of stilt-like protuberances that rise above the ground to the base of the trunk. They make it look like the palm might be about to stride off with a spidery gait across the swampy terrain. So much so, that it was suggested in 1980 that the palm did indeed "walk"—growing new roots when it wanted to shift position, letting the ones "behind" it rot away, allowing it to literally cover ground in slow strides.[14] There is no evidence for this idea, however.[15] The walking palm does not actually perambulate. It is much more likely that the rooted "legs" allow the palm its surprising proportions, supporting the spindly trunk in rapidly attaining great heights to reach the light without needing to invest in architectural girth. They also might allow the palm to root itself on ground that is very uneven and covered in logs and trunks. Growing these roots solves a significant problem for the palm: getting to the light quickly in an overcrowded forest without having to go through the achingly slow process of building a dense, thick trunk.

Some plant adaptations solve the problem of getting sustenance in novel ways. Most plants synthesise molecules such as glucose using energy from the sun through photosynthesis. They are self-sufficient, to a degree, though they usually have intimate relationships with fungi at their roots. These allow plants to absorb other nutrients from the soil to complement their diet of sunlight. A small number of plants, though, have bypassed this arrangement altogether, managing to have their cake and eat it, as it were. They tap into the fruitful mycorrhizal network between the fungi and tree roots to draw off resources, but do not contribute by doing any photosynthesis on their own. One of these plants was discovered in 2015 by a team led by Kenji Suetsugu at Hakubi Center for Advanced Research, Kyoto University, on the Japanese subtropical island of Yakushima. It usually remains underground, sending up deep red stems only a couple of inches tall bearing flower buds on the rare occasions that it wants to reproduce. The rest of the time it lies low, infiltrating the symbiotic connections between the ancient cedar trees on the island and their fungal networks, siphoning off nutrients. The aptly named *Sciaphila yakushimensis* (*sciaphilic* meaning "shade loving') are essentially guerrilla plants, parasites that manage to bypass the usual burden of photosynthetic toil.[16]

Plants also have adaptations that allow them to play mind tricks on creatures around them. Tomato plants, for example, produce certain chemicals when they are attacked by plant-eaters such as caterpillars.[17] John Orrock and colleagues from the Department of Integrative Biology at the University of Wisconsin tested how exactly these chemicals protected the tomato plants.[18] They found that they have a grisly effect on plant predators: they turn them into cannibals. The chemicals both make the plants taste terrible to the caterpillars and warn nearby plants to start producing them, too. The hungry insects then start to attack other insects instead of the leaves of the tomato plant. This has the dual benefit that the insects sate their appetites carnivorously rather than with their usual vegetarian fare, and reduces the numbers of plant-eating insects overall.

Astounding as this cross-species mind control is, though, it is still "only" an adaptation. Plants suffering the depredations of insect attack respond by releasing these cannibalism-inducing chemicals, inculcated into the genes over evolutionary time as a result of a bitter herbivore–plant arms race. The insects come with their scything mouth parts and digestive abilities; the plants respond with cellular armaments and chemical weapons. As with the walking palms or robber *Sciaphila yakushimensis*, no cognition is needed.

Breathless anticipation

There are some plant behaviours that appear on the surface to be simple adaptive responses, but which, when you probe further, are much more complex. Being able to anticipate changes that might happen in their environments, such as rain or when the sun will rise, allows plants to prepare for them and be ready to maximise these opportunities, which pays off in the long term. In the tropics of Africa, for example, plants deck themselves with foliage *before* the rains arrive to ensure they capitalise on the growth season ahead.[19] We have seen the light-seeking growth of beans and other climbing plants. But some plants track a moving light source—the sun—through the day. They are sun-worshippers, heliotropic plants. Their leaves and shoots dynamically follow the sun through the sky over the course of the day with incredible accuracy. Young sunflowers do this by turning their heads to follow the sun east to west, deviating less than 15 degrees either ahead or behind. This maximises the sunlight falling on the flowers and, as a consequence, the number of pollinators attracted to them.* [20] Now, it might seem a simple task for a plant to track the sun from the direction of the light hitting it—until we know that plants can accurately follow the sun even when it's cloudy. If you rotate a young sunflower 180 degrees during the night, it will take a few days to reorientate its movements to the new angle of the sun relative to its bloom. The plants are not just responding to what is happening around them, they might have an internal model of what the sun is going to do that guides their movements.

Things get even more mysterious when we look at what plants do at night. Many of these sun-worshippers, including young sunflowers, reorientate their leaves or blooms at night to face where the sun will rise. It is not simply a retracing of the movement of the day; it happens at double the pace, even in the absence of any cues from the sun the night before. Remember Cornish mallow or *Lavatera*, the little plant that can predict where the sun will come up and turns its leaves to face it in preparation, managing to do so for a few days even if deprived of any sunlight? This behaviour is adaptive, maximising the sunlight that the leaves can soak up during a day. It is also predictive: the leaves don't turn in response to the sun, they are ready in *anticipation* of sunrise.

---

* Many others track the sun, too, some of them as familiar as cotton, together with many other members of her family (Malvaceae). Other well-known sun-chasers are soybean or alfalfa. Once the sunflowers are mature and open, they are transfixed, facing eastwards to optimise the sunlight hitting their blooms.

Lavatera manages this in part through a delayed-response mechanism. It uses the starch granules that accumulate as a result of photosynthesis to "mark" the position of the sun. As the plant is exposed to sunlight, photosynthesis causes sugars to build up which are converted into starch granules. In the morning, when the light hits the plant from one side, these granules are deposited on one side of the stem. During the day, when the light is overhead, they build up evenly. Overnight, when there is no photosynthesis occurring, the starch is broken down for energy. But because there was more starch on the side of the plant the sun hit at sunrise, more granules remain by the end of the night. These affect how the water content of the cells is regulated on each side of the stem, causing the stem to bend towards sunrise even before the sun has come up.[21]

The reason why *Lavatera* and other plants reorientate towards sunrise at night is because getting a head start is always a good idea. Maximising the photosynthesis they can carry out in a day gives them a big advantage, especially in regions of the world where sunlight is not all that plentiful. A bit like students preparing for lessons and arriving punctually at school, these plants can both perform the metabolic reactions needed to prepare for photosynthesis, and absorb as much sunlight as possible throughout the daylight hours. Predicting where and when the sun will rise must mean that, to a certain degree, plants can model their environments internally in some way. The set of mechanisms that allows flowers to track the sun in shaded conditions are related to the circadian rhythm, an internal model of the cyclical changes *outside* the plant that governs the timing of changes *inside* the plant. This is entrained by key cues such as light and temperature to keep the internal clock on track. Keeping accurate time is key: the clock keeps the plant in sync with what is going on around it, not only in response to changes, but pre-empting and preparing for them. Plants that can run their internal functions and interactions with the environment in concert with the changes going on outside them fare much better than those which have had their circadian rhythm genetically knocked out, and live out of kilter with these cycles.[22]

Why might predicting changes in the environment, and when they will occur, be so important to plants? If we can answer questions such as this, it might make us more receptive to what plants are capable of, because it will show us that they cannot but have abilities beyond the reactive. There is more than one way to think about it: my colleague Ariel Novoplansky, plant ecologist at Ben-Gurion University of the Negev, Israel, and I each emphasise different aspects. My perspective focuses on the rapidly moving complexity of the *environment*. I argue that plants cannot afford to make poor decisions in a fast-paced biological economy. For their behaviour to be at all adaptive, they must take the future into account, because things move quickly—something that might be said for mobile organisms too. They have to anticipate if they are to be fit for survival in the environment a few hours hence, or tomorrow, or weeks away. So, roots have to grow in a way that predicts where resources will be; shoots turn, grow, bud and flower guided by predictions about where there will be sunlight, how the seasons will change, or whether there will be enough minerals and nutrients to support growth in the future. Flowers can even time their pollen production and presentation to fit the times when pollinators are likely to visit by extrapolating from past experience.[23]

Ariel's perspective highlights the slowness of the *plants'* pace of life. He argues that plants do everything so slowly that they can't try again if they get it wrong. They only get one shot at making the right choice, so they had best get it right the first time. Mobile animals, according to his approach, don't have such pressure. If an animal makes a mistake about the direction it goes in or where it looks for food, it can trail back and try again fairly quickly. If a plant invests a lot of energy growing in the wrong direction, and finds no nutrients, water or light when it gets there, it is in deep trouble.[24] So the information that guides plants' growth and behaviour often needs to be information about the future. It must be *anticipatory* growth if it is going to be any good at all.

Both of these approaches arrive at the same conclusion, whether you emphasise the speed of the changing environment or the slowness of the changes which plants make: plants need to predict. We would not be surprised if plants evolved to respond as early as possible to environmental changes. So, why wouldn't they use this information to learn and make predictions, just as animals can?

Dealing with complexity

Light—where and when it will appear—is only one of the many things that plants care about. They live in complex worlds, just as animals do. They need many, nuanced ways of collecting information and using it to guide what they do. Adaptations, well-honed automatic responses to the environment, allow for simple and effective ways of dealing with common problems: tapping into resource-filled hosts, growing towards light, deterring herbivores, keeping upright. But they don't give rise to the flexible capacities that will allow plants to fine-tune their actions to optimise for multiple, dynamic aspects of their environments. For this to happen, numerous sources of information have to be collected together, integrated, and used to guide behaviours, which can be incredibly flexible by virtue of how plastic plants are in the way they grow and develop.[25]

These two key aspects, integration and flexibility, are worth pausing to look at more closely. We tend to think of plants as organisms that grow towards or away from things—towards the light, away from gravity, towards water. But there are so many more aspects of the living and non-living environment that plants have been shown to respond to in experiments. They respond to five different parts of the light spectrum, as well as day length and changes of season. They also respond to humidity, vibrations, salt levels, variation in nutrient availability over time, the micro-organisms in the soil, competition from neighbours, being eaten, wind, temperature and many others.[26] Plants are juggling the demands of these numerous different factors all the time— and sometimes they will have to triage between them. You can't optimise for everything, especially in a constantly changing and complex environment alongside other organisms which are also trying to make the best living they can.

There is a clear case to be made that these trade-offs are being played out on the underside of the leaves of plants all the time. Leaves are not just sun-absorbers, they also have tiny pores called stomata, generally found on their undersides, which allow gases and water vapour in and out of the leaf. They can open or close depending on the needs of the plant. Most importantly, on a sunny day, the stomata keep the leaf cells supplied with carbon dioxide, which is a vital raw material for photosynthesis. There is a catch-22 though: on hot sunny days when the plant needs more carbon dioxide, the sun hitting the leaves will also make more of the water in the plant evaporate. So, to keep the stomata wide open in order to let more carbon dioxide in, the plant also has to allow more water vapour out. If there is plenty of water in the soil to draw up from the roots, this is no problem, but if conditions are dry, dehydration is a serious risk. To balance these needs, the stomata are sensitive to both the levels of carbon dioxide in the leaf and the stress signals from the root, which are conducted via a stress hormone chemical called abscisic acid.[27] The different levels of these signals fine-tune how open the stomata are to provide as much carbon dioxide as possible when it's needed without risking the plant becoming perilously wilted. The signalling can even form a kind of memory. If plants live through a period of drought, to control the future opening of the stomata they use a signalling molecule which is also found in animals. GABA, or g-aminobutyric acid if you want to use its full name, remains in their cells as a reminder of the intensity of the dry spell.[28] Even at the level of individual cells, the multiple demands of plants' lives have to be carefully balanced.

From the perspective of the whole plant, resources are limited, so they have to be used wisely. Plants continually keep tabs on the numerous aspects of their surroundings and bring this information together to guide their physical growth and physiological responses in ways that will give them the best chance of thriving. Some of these activities equate to behaviour that we might think of as very animal. They demonstrate self-recognition and territory-guarding, and plants can make internal maps of their surrounding soil to guide root growth, seeking rich patches and avoiding obstacles before they have even encountered them.[29] These abilities rely in part on the ability to detect where all of their relative body parts are, similar to the "proprioception," or sense of where all their body parts are in space, that animals have.[30]

In order to bring all of this information together, plants need to communicate between the different realms they inhabit, most obviously their "above-ground" and "below-ground" parts. There has to be cross-talk between the roots and shoots, which integrates the signals the plant is continually collecting into a more complete picture of the world around it. The plant reaches out with the microscopic tips of its root hairs, or the growing ends of its shoots, sensing everything it can at the boundaries of its furthest reach. The effects of this information have to be communicated to the rest of the plant. Only then can the plant balance all of the competing demands on its resources and respond effectively. For example, plants decide where to invest in their growth depending on what their neighbours are doing. If they grow too close to one another, they block each other's access to light. A plant growing in a crowd therefore needs to get up high as fast as possible to keep access to light, supercharging its shoot growth over that of its roots. How does it tell? One sign is physical touch, the contact made with the leaves and stems of its neighbours. This is communicated throughout the plant, right down to the roots. A plant that makes above-ground contact with its neighbour will produce chemicals from its roots that communicate the crowded situation to the plants nearby.

We can see how information about how crowded the neighbourhood is affects plants' decisions by offering them a choice: a fork in the road on their growing route, to see which direction they pick. One study tested how young maize plants grew their roots in a "Y-maze," an inverted forked container with different solutions at the bottom of each fork. On one side was a solution in which plants were growing that had been touched on their leaves, simulating crowded growing conditions. On the other side, a solution in which "untouched" plants were grown. Almost all the seedlings chose to grow into the "untouched" solution. It seems that there was something in the "touched" solution which made that avenue of growth far less appealing. Plants given no choice, placed in the "touched" solution, also invested heavily in their shoot growth compared to their roots, suggesting that they felt the need to get ahead of the competition.[31]

Plants communicate in this way all the time; they can also affect the flowering activity of their neighbours. If *Brassica rapa* plants are given artificially long days, for example, they will flower more quickly and put less energy into growing storage organs in their roots. If they are grown next to plants kept in conditions imitating short days, a long way from flowering and investing heavily in storing energy in vegetative organs, something strange happens. The short-day plants will start to flower earlier and neglect their storage organs. The fair-weather plants seem to communicate the apparent halcyon days to the short-day plants through root chemicals, and entice them to act accordingly, even in the absence of external signs.[32] The cross-talk at the roots affects what happens in the above-world parts of the plant, allowing plants to integrate the information they are collecting throughout their bodies and devise an overall strategy.

Growing with the flow

Just as looking at one aspect of the plant's environment and how the plant responds to it will give us a woefully simplistic picture of what the plant is doing, we also can't assume that how they respond in one set of circumstances will hold for all. Plants may be rooted in one spot, but that means that they have to be better at dealing with the changes around them. They can't move to greener pastures like animals can, they have to take things as they come. They can't run away from predators or parasites, they have to handle them as they arrive.[33] Given all the different things that plants monitor and anticipate in their environments, they have to be incredibly flexible in all sorts of ways—in the way they grow, how they time things like reproduction, or how they defend themselves. They can override the usual tendency of roots to follow gravity and head downwards if they sense barriers in the way, they can mount defences against drought or frost when exposed to drier conditions or mild cold, or they can change the way their stems grow based on experiences while they were previously dormant. They can turn their leaves to the sun when the soil is moist, and away from the sun when the soil is dry— making trade-off decisions as animals do.

When a plant "behaves" it does one of several things.[34] First, it can make irreversible changes in its growth patterns for long-term, slow "movement," as climbing beans do. Second, it can make reversible changes in the water content of different cells in order to perform short-term movements, as with the Venus flytrap or leaf stomata. It can make specialist organs and tissues such as flowers. Or it can change the chemicals that it produces— altering its physiology, as we saw in the tomato plants. We have to bear all of these in mind when looking at what plants do—they are not the same as animal "behaviours."[35] Behaviour does not usually change an animal's growth pattern significantly from what is genetically encoded within that animal. But the decisions a plant makes in its growth and movement—which direction to go in, when to branch, clasp or flower—are what determine the shape and form of the plant. The rigidity of plant cells is counterbalanced by the indeterminate nature of the forms they can take. In scientific parlance this is called *phenotypic plasticity*:[36] the phenotype being everything you could physically observe about an organism. An animal will largely be physically the same in whatever situation it develops. But an individual plant in one environment is not at all the same plant—physically or behaviourally—as it would be were it somewhere else, in different circumstances. It is the plant's complex engagement with its environment that shapes what emerges. And, as we shall see with the following examples, what might underlie this plasticity is something that could be called cognition: adaptive, flexible, anticipatory, goal-directed behaviour.[37]

Through the grapevine

There is a long list of plant behaviours that we suspect are underpinned by cognitive processes, from learning and memory to competitive, risk-sensitive behaviours, and even numerical abilities.[38] Memory itself, for example, has to be learned, and is a vital ability for survival. Plants show numerous examples of responses to something they have encountered before that have been primed by their earlier experiences. They are quicker to defend against herbivores or parasites if they've been previously attacked. Changes in temperature or chemical environments can have effects that trickle down for five to twelve generations.[39]

Not least of all the complex behaviours that might have cognitive underpinnings is the constant communication that plants are having with those around them and even with other species, through multiple channels. Plants speak the silent language of scents. They do it through their leaves, shoots and roots, and of course through their flowers and fruits; trees discharge them into the open air even through their barks. Virtually all plants have mastered the tricks of chemical talk, synthesising and releasing into the air many different volatiles (volatile organic compounds, or VOCs) throughout their whole bodies for a number of purposes. VOCs provide valuable information in the form of sophisticated blends of terpenoids (mostly isoprene), but also benzenoids and other compounds. We may picture each volatile itself as a building block in the vocabulary of plants, with "words" being made up of many different organic compounds, a bit like Lego sculptures.[40] All in all, plant communication relies on an ample lexicon with a size of over 1,700 different volatile cocktails.[41]

Plant behaviour can change dramatically as a result of the messages being exchanged.[42] Subtleties can make a world of difference when it comes to conveying a particular message. For example, the characteristic "green odour" that you can smell from freshly mown grass is a result of wounding the grass leaves. VOC distress signals warn other grasses nearby that danger is at hand and alert them to mount defences. Sometimes plants of different species warn each other: for example, sagebrush (*Artemisia tridentata*) and wild tobacco (*Nicotiana attenuata*). A tobacco plant within airborne reach of a damaged specimen of sagebrush is less affected by the attention of herbivores. The release of a number of VOCs by the good shrubby Samaritan alerts the tobacco plant, which switches on the machinery to produce repellents.[43] Early warnings and fast communication can make all the difference.

These messages cross the boundaries of the plant world, too. We've already seen the way that tomato plants produce chemicals to muddle the brains of herbivores feeding on them and turn them into cannibals. Other plants and trees under attack will recruit their own "bodyguards." They use airborne chemicals to attract predatory insects that happen to have an appetite for the herbivore threatening them. Undamaged lima bean specimens (*Phaseolus lunatus L.*) help themselves to terpenoids to recruit predatory mites (*Phytoseiulus persimilis*) that hold the annoying spider mite (*Tetranychus urticae*) at bay.[44] Others have nectar-secreting glands that lure in ants with the promise of sugars. The ants then act as sentinels to keep herbivores away.[45]

This constant network of communication suggests that plants have some kind of social intelligence. And one of the fundamental elements of social intelligence in animals is recognising your relatives, because they are likely to work with you rather than against you: you share genetic material, after all. Aside from VOCs emanating through the air, plants can use the chemicals exuded at their roots to speak to one another and identify each other. Remember the maize seedlings with their Y-shaped mazes? Well, other plants have been shown to be more aggressive when fighting for underground resources with different species compared to their own. American sea-rocket (*Cakile edentula*) plants grown in a pot with strangers will produce a much larger root mass than when they are growing alongside relatives, making them more likely to win in the foraging race.[46] Above ground, plants can sometimes "see" whether others are relatives or not. The *Arabidopsis* plant, used widely as a model organism in biology, seems to use the unique wavelength profiles of the light reflected by its neighbours to tell whether they are relatives or not. When they are grown together with kin, these plants will produce much more seed than when growing with strangers. It's possible that living among family makes life easier, so plants can afford to invest more in reproducing.[47]

Plants can even assess the risks of the choices that they make, which is important when resources are limited. Plant growth tends to be enhanced in richer soil patches, but roots do not just care for water and nutrients. Things are not as simple as following the command to, say, "grow more under low concentration of nitrogen" (an element essential for plant growth). Optimising foraging in search of richer patches in the face of predation and competition requires plants to keep a constant eye on a number of parameters as they fluctuate in real time.[48] After judicious cost-benefit analyses, plants decide where to invest their precious metabolic resources.[49] Peas, for example, can be bearish or bullish in their root growth depending on the circumstances. In one experiment, pea plants were grown with their roots split between two containers, making it possible to see how they make risk judgements about where to grow. In one container, plants were offered a constant level of nutrients, in the other variable levels were made available. When sufficient resources were provided to the plants with a constant nutrient supply, they didn't bother growing roots into the container where resources varied.[50] They were bearish about where they invested their energies, going for the safe, reliable option. If the nutrients in the constant container were too sparse, however, the plants took a punt, growing roots into the variable container, being bullish by necessity. One could even argue that the plants were making *decisions* about what their strategy should be, based on some kind of assessment of how necessary it was to take risks.[51]

Peas remember

One of the most exciting areas of research into plant intelligence that is just beginning to bloom reveals that plants have the ability to both learn and remember. Individual plants can acquire new information about the environment, retain this information, and use it to guide their future behaviour. It is not an entirely new idea. The shy mimosa plant we met in the Introduction has long been a focus for botanists exploring how plants might learn. Its sensitivity and folding response captivated botanists in the eighteenth century, including R. L. Desfontaines.[52] He carried out an experiment which was easily replicated, placing a mimosa in a moving cart. At first, the shaking of the cart caused the plant to fold up its leaves, but after a while, the plant opened them again, apparently used to the movement. If the cart stopped still for a while and then began moving again, the resumed shaking would cause the plant to fold up hurriedly once more before it again got used to the movement and opened its leaves.[53] Later, in 1873, Wilhem Pfeffer showed that the mimosa would stop responding to being touched over time if it was prodded too often.[54] This is a very simple form of learning, called *habituation*, where a stimulus happens so frequently and inconsequentially that the reaction to it is blunted and is eventually ignored.

When Monica Gagliano, author of *Thus Spoke the Plant*, investigated mimosa's habituation tendencies with her colleagues at the University of Western Australia in 2014, they found it had two fascinating and complex aspects.[55] First, when a plant was in an environment where there wasn't much light, it was much quicker to habituate and stop folding up in response to touch than plants in situations where light was plentiful. The downside of folding was greater when it might cost the plant valuable time doing photosynthesis, so being hyper-sensitive to the risk of a nibbling herbivore became relatively less important. When light was plentiful, the plant could afford to be hypervigilant and avoid anything that might signal predation. Second, the mimosa's habituation was not short-lived: it could last for up to twenty-eight days. Mimosas seem to have long-term memory.[56]

Monica and her team moved on to investigate other, more sophisticated types of learning in plants, ones we tend to think that only animals are capable of. "Classical conditioning" or "Pavlovian conditioning" is where a subject learns to respond to a neutral stimulus that would usually not cause a response, when it is paired enough times during "training" with a stimulus the subject would naturally respond to. Think of Pavlov's dogs salivating when they hear a bell, because the bell has so often been paired with food. No food need be forthcoming for them to respond to if they have been trained thoroughly enough. Monica and her team found that pea plants can learn in just this same way. They placed pea plants in the same Y-maze we saw for maize roots earlier on, right way up to give the pea shoots a choice of where to grow.[57] The seedlings were presented with photosynthesis-fuelling blue light on one side of the Y as the "reward" towards which they would grow. When there was no blue light, they chose the direction from which blue light had last come. But, when trained for several days by preceding the blue light with air movement from a small fan, to which the seedlings previously had no response but could sense, something fascinating happened. Seedlings could be drawn to deviate from their natural response—to grow towards the branch of the Y-maze from which blue light last appeared—if the fan was sensed from the *other* side. The fan had, it seems, come to mean "dinner" to the well-trained pea seedlings.[58]

Darwin himself made an observational case for learning in germinating plants. He observed that the young leaves of seedlings responded differently to light depending on their past exposure to it.[59] There are very significant advantages that plants might gain by having flexible behaviour shaped by their individual experiences. So it might not be all that surprising to find that they are capable of something seen hitherto as the preserve of animal behaviour. Think of a plant seeking out nutrients in the soil, for example. Investing in root growth is a costly business, so if plants can work out where it is likely to be valuable to do so, they will be more likely to capitalise on what is available to them. Latzel and colleagues managed to "teach" wild strawberries (*Fragaria vesca*) to associate light intensity with the availability of nutrients in the soil. Some plants were trained to associate high light intensity with rich soil patches, and others to associate low light levels with rich soil patches. The training held when the plants were grown in conditions where the two things were not linked. The "light" plants grew roots heavily where light intensity was high, and the "dark" plants invested in roots where there was shade, even if the soil didn't vary in richness.[60] Individual strawberries were able, within their own lifetimes, to get canny about how to find the things they needed in their environment, connecting new dots, even though the clues used in the experiment never would have occurred naturally.

Despite recent research, resistance to acknowledging the very concept of plant learning remains fierce. According to conventional wisdom, while animals learn, plants evolve adaptations. We are much more comfortable accepting that a mollusc or fish might learn than a plant. In his *New Yorker* report mentioned in Chapter Two, Michael Pollan shared parts of his conversation with plant physiologist Lincoln Taiz. As Taiz insisted, in referring to recent research on mimosa and its habituation, the words "habituation" or "desensitization" would be more appropriate than "learning." And yet, according to the *Penguin Dictionary of Psychology*, habituation refers to "the gradual elimination of superfluous activity in learning" and to "a form of non-associative learning whereby each repetition of a stimulus results in a progressively diminished response."

I will be the first to admit that there is some patchiness in the research supporting plant learning. Our work at MINT Lab has not yet replicated the findings of Monica and her colleagues on Pavlovian classical conditioning, but we are in the process of investigating it. Alongside studies such as Monica's that suggest that plants are capable of Pavlovian classical conditioning, there are a number which seem to show that it is not present or that the results are unclear.[61] More recently, *Arabidopsis thaliana* has been reported to exhibit conditioning in response to heat stress, but this, too, awaits independent replication.[62]

Training plants and testing their learning in a robust way is not so easy when we are only just beginning to understand what their subjective worlds are like. Not only that, but as we've seen, plants are deeply shaped by their surroundings: can we *really* test what plants are capable of in sterile laboratory settings? It might be that we need to find ways of doing such work in rich ecological contexts where plants are expressing themselves fully. The possibilities from our research currently underway are extremely exciting.

PART II

THE SCIENCE OF PLANT INTELLIGENCE

Tell me why the stars do shine
Tell me why the ivy climbs

Fred Mower and Roy L. Burtch, "Why I Love You"

CHAPTER FOUR

PHYTONERVOUS SYSTEMS

In June 2018, I waited in line for tickets at the New York Botanical Garden. To pass the time, I flicked through Sir J. C. Bose's *The Nervous Mechanism of Plants*, which has become a lodestone for me. One sentence in particular caught my eye: "No form of ganglion however has ever been observed in plants but it is not impossible that the physiological facts may one day receive histological verification."[1]

Holding the book in my hands, I mulled over plants and ganglions as the queue moved slowly forwards. Bose's remark might prove prophetic.[2] Indeed, there are no observable brain-like structures in plants, but even when Bose was writing, plant science was throwing up surprising facts about plants' interior workings. They fired the imaginations of some biologists. It seemed that plants might have far more complex interior worlds than we knew, we just couldn't *see* it. Could plant science end up revealing a system in plants akin to the nervous system of animals? If not grey, do plants have their own "green matter"?

As I passed through the ticket booths to the gardens and saw inside, I couldn't believe my luck. A curator had put up a small but jewel-like exhibition, titled *Visions of Hawai'i*, of paintings of landscapes and plants of the Hawaiian islands by the American modernist Georgia O'Keeffe. One work named *Papaya Tree* immediately arrested my attention.[3] I had just been looking at the plate of a micrograph from the main stem of a *Papaya* in Bose's book. It showed the distribution of tubular "vascular" tissue, which carries water and nutrient-rich sap through the plant. I had bookmarked it with a Post-it note. The painting seemed to animate the ideas I had absorbed from the book I was holding and set them in motion in my imagination. As I looked at the painting, it felt like I was seeing *into* the tissues of the tree, the fine mesh-like network of tubes running up and down its stem, which might be the very basis of the internal communication systems in plants which Bose had written about: "a system of nerves enables the plant to act as a single organised whole." I had both the microscopic detail and the astounding brush-stroke depiction of the "organised whole" before me in that moment. The organism thriving in its environment and the physiological secrets that allowed it to do so.

Speaking of nervous systems in organisms that lack nerves might seem somewhat misguided. But we have known for a long time that plants are capable of electrical transmission through their tissues. Even 150 years ago, Charles Darwin suspected that some form of electrochemical communication was behind the responses of carnivorous plants such as the Venus flytrap, which he called a "most animal-like plant."[4] Since he had no way to measure electric currents in plants, he shared his insights with physiologist Sir John Burdon-Sanderson, at University College London, who was able to measure a voltage difference between the upper and lower surfaces of leaves. Later, in the 1930s, inserting microelectrodes into the giant cells of the freshwater algae *Chara* and *Nitella* revealed how this cellular excitability was produced, when the first nerve-like electrical impulse was recorded.[5] Bose also investigated the electrophysiology of plants in detail, working particularly on mimosa. He showed that an electric impulse triggered the folding of a leaf via the "transmission of excitation." These and other studies revealed the serious need to consider that electrical processes—akin to those in animals—partly underlie plants' internal signalling and ultimately their adaptive behavioural responses.

It is extremely easy these days to demonstrate electrical activity in plants. If you apply some conductive gel to the surface of a Venus flytrap leaf, and measure the voltage changes across its surface with an electrode, you will see that by brushing the sensory hairs of its surface an electrical signal is generated, an impulse that spreads quickly over the entire trap, causing it to close.* However, electric signalling is not just a feature of those plant species capable of fast, surprising movements. It is found

* Buy yourself a Venus flytrap, some electrode wire and an amplifier. Download some spike recording software (Backyard Brains: https://backyardbrains.com will do). Apply conductive gel between the outside of a trap and the wire. After repeated stimulation of the trap, an electric impulse will spread throughout the plant body, and it will close. You will get to see the whole thing on a screen.

virtually everywhere. Plants all regulate their physiological processes in the same way. Light, gravity and touch can trigger an electrical response; the same goes for sudden changes in temperature, water resources, or salt stress. Pathogens, herbicides and other chemical substances, or cutting, wounding and burning, can make a plant fire electrically too. So can an animal's bite or the removal of the plant's leaves or fruit. Even when a hibiscus flower is visited by pollinating insects, the intimate act of pollination triggers signals which result in an increase of the respiration rate in the ovary sitting at the base of the flower.[6]

When we think of electrical communication in organisms, we think of the rapid speed of animals' nervous transmission, but plants have evolved conductive devices for their own unique purposes. They use the signalling capacities of their particular kinds of networked cells to coordinate their systems. That we are reluctant to see this communication because it does not happen in nerves shows the limited nature of our own ideas. If we go back to basics and ask what it is that neurones actually *do*, we could say they generate and share electrical charges. They chat to each other in the form of spikes or *action potentials* firing and passing along cells and between them. What makes an action potential, according to the *Oxford English Dictionary*, is "the change in electrical potential associated with the passage of an impulse" along a cell membrane. The movement of voltage changes along membranes is the essence of nervous communication. But we have known for a long time that this is not exclusively a property of the nervous system, even in animals. Animals' muscular cells can spread electric waves throughout whole organs. Think of the contractions of your heart as an electric impulse spreads across your cardiac muscle tissue. So there is no reason to deny that plants might use electrical signalling just because they lack neurones.

Plant cells lack the neuronal structures that animals have for transmitting electrical impulses. How then can information travel from one plant cell to another throughout the plant? In the absence of nerves, electrical signals can travel along the vascular system, the transport network formed by bundles of tubes that stretch from root to shoot. This is made up of two types of vessel: the xylem, which carries water upwards through the plant, and the phloem, which transports dissolved substances such as sugars. Think of the vascular system as akin to the nervous system of animals, acting as a freeway that conducts electrical information over short and long distances. Just as the nerves in an animal resemble electrical wiring that can conduct electrical signals, the vascular system is like a green cable that carries news throughout the plant in the form of electrical signals for the purpose of controlling and coordinating plant functions.[7]

Just as animal nerves do, the electrical circuitry in plants runs on a variety of electrical firing events, among them action potentials. And yet generally speaking there is little mention of action potentials in plants, even in Lincoln Taiz and Eduardo Zeiger's classic *Plant Physiology*, a work of reference for students and researchers alike. Yet today we know that plant action potentials have a very similar firing pattern to those in animals, travelling over long distances along the vascular system.[8] As early as 1963, pumpkins were reported as firing spikes of voltage.[9] These signals allow plants to gather information and coordinate their different structures in order to achieve intelligent goals. The "neurological" system of plants is an integrated, excitable network interlinked by numerous, irregularly distributed cross-links. Bose identified up to twenty layers of vascular tissue, nesting like Russian dolls, one within the other. Stem layers connect radially, forming the "complex net-like structure" that Bose observed in papayas—Georgia O'Keeffe's model included.

While both the vascular bundles of plants and the nerves of animals conduct electrical signals, animal nervous systems organise signals differently. They evolved to coordinate free-moving behaviour, carrying a signal from point A to point B in a precise, targeted way. Plants must also coordinate their behaviours in response to a variety of signals, such as herbivore attacks, variations in light and temperature, mechanical stimulation or salt stress, among many, many others. But the behaviours of plants are usually slower and more generalised. They can result, for example, in changes in photosynthesis or respiration, or in gene expression. Differences aside, is the highly branched, excitable nerve-like system of a papaya plant the functional equivalent of a hierarchically organised but diffused brain? We cannot answer this question just yet, but it is a very exciting prospect.[10]

Green neurochemicals

Evolution has left us a pretty sizeable clue to the early origin of internal signalling, buried deep in the evolutionary history of both plants and animals. Signalling between cells arose in the earliest multicellular organisms before any kind of "neurological" structures developed in animals at all. And this can result in abilities that are nothing short of surprising. Slime moulds, for example, are a heterogenous group of organisms that used to be seen as fungi but now inhabit the *Protista* kingdom. They are otherwise known as "blobs" because they form large mass-cells with many nuclei and very wide skill sets. They can solve everything from maze problems to algorithmic puzzles, and remember molecular likes and dislikes, because of the communication between individual moulds that fuse together.[11] If single-celled organisms need to communicate, this need only escalates quickly when you get into multicellular life, if there is to be any useful division of labour or collective response to the environment. Not only that, many of the signalling molecules that are involved in these cellular conversations are still present in both plants and animals, inherited from those early interactions between the cells of their common ancestors.

The essential chemical and electrical messages are the same across life: guiding the growth of an oak tree, the leaf-folding of mimosa, the rhythmic contraction of a jellyfish's mantle (an animal with no nervous system to speak of) or the astounding sprint of a cheetah, the fastest land mammal on Earth. In the same way that plant and animal cells share common ancestors, have many similar structures, use the same mechanisms for gene expression, and have similar metabolisms, they also speak with some of the same languages. This should not be a surprise. If the rapid and long-distance nature of electrical signalling is worth the energy cost to animals, why not to plants? If it weren't, it would have been cut as evolutionary deadweight from plants' cellular repertoires a very long time ago: electrical signalling is not biologically cheap, so tissues that use it are only maintained if they are doing something beneficial over evolutionary time. We know this, but have still tried to fiddle with the labels to avoid admitting to a "neurobiology" in plants. The chemicals involved in the transmission of electrical signals between neurones in animals, neurotransmitters, are called *biomediators* in plants. Yet the chemicals such as acetylcholine, catecholamines, histamines, serotonin, dopamine, melatonin, glutamate and GABA found in plants are the same molecules as are produced by animals.[12]

Take GABA, now thought of as a key component of the animal nervous system. This molecule is an amino acid that decreases the receptivity of neurone membranes to being excited by electrical signals. It was assimilated into the recognised animal biochemical arsenal in the 1950s, when it was discovered that GABA played a key role in mammal brains, and also in crayfish. But before this, GABA was not seen as an "animal" molecule; it had first been synthesised in 1883, and was thought of as a metabolic product in plants and fungi. While animal studies focus on its role in neurones, in plants GABA has been studied primarily for its metabolic roles (such as pH regulation). GABA's importance in plant signalling has garnered interest over the past couple of decades, though.[13] In fact, GABA receptors have been found in plants, confirming its role as a signalling molecule: not only does the plant produce it, its cells can detect and be affected by it. One of its roles, stimulating rapid defences against insects and other damage, is beginning to be understood.[14] Not only is GABA produced by plants, there is no doubt that it has a function. Some plant physiologists assert that there is "no evidence that it functions as a signalling molecule in plants."[15] But assertions cannot change the molecular reality.

Another molecule which is important in memory formation in animals also does something extraordinary to leaves when they are wounded. Glutamate is an amino acid like GABA, and we have known for a long time that it too is produced by plants. But in 2018, a team led by Simon Gilroy at the University of Wisconsin-Madison revealed just how important glutamate is for them.[16] The team genetically engineered an *Arabidopsis* such that it contained a molecule that flashed brightly when calcium levels increased in the cells. When they wounded these plants with a blade, they saw waves of light ripple away from the wound site, showing the flow of calcium release passing to undamaged areas. And when they investigated further, they found that glutamate was responsible, stimulating a calcium-based wave of electrical activity, signalling to the cells to go into defensive mode. This "mammalian" neurotransmitter seems also to carry rapid internal distress signals in plants, not unlike the way that neurotransmitters operate in animals. In fact, the genes coding for glutamate receptors in animals are extremely similar to those in plants: *Arabidopsis* has twenty of them. Glutamate may be involved in other things, too: shaping responses to light, guiding root growth and sensing where nitrogen is to be found in the soil.[17] Molecules like GABA and glutamate act as signals between cells in both animals and plants, guiding what cells do, how they grow and develop. This is especially important for plant behaviour, which is based on cell growth and development.[18]

So we have begun to understand the basis of electrical signals in plants, but what of the potential for learning that we touched on in the previous chapter with Monica Gagliano's peas? In animals we have known for a long time that stimuli can become linked with one another, so that a stimulus which previously didn't elicit a response can come to elicit the response spontaneously created by another stimulus, in the process otherwise known as "classical conditioning." We didn't know what the neurological basis of classical conditioning in animals was until recently. In 2020, a team led by Sebastian Haesler at the Flanders Institute for Biotechnology found that mice given a completely new stimulus—in this case a new smell—when they were fed, tended to associate that smell with food more quickly than they did a smell they were already accustomed to. They suspected that the novel smell activated the dopamine system, the same one that gets triggered when we see a notification on a social media app and keeps us glued to our phones. If they gave the mice dopamine blockers, however, the effect of the new smell on learning speed was almost as slow as with familiar smells. Other researchers have found that the neurones firing in response to more than one stimulus become more ordered and synchronised in their behaviour over time. The stimuli, in effect, become neurologically associated with each other. The "Pavlovian response" might be down to dopamine and coordinated neurone responses.[19]

The fuller understanding of it we have gained in animals might show us avenues to explore when investigating how learning might work in plants too. Dopamine is found in plants in quite high concentrations. If we look back at Monica Gagliano's experiments with her pea plants learning to navigate their Y-mazes, we can conduct a thought experiment. If synchronised neurones firing in response to sensory stimuli, excited by dopamine, cause learning in animals, then could the same not apply to plants? If electrical signals are sent in response to sensory stimuli in plants, then perhaps they too can have coordinated responses which cause them to learn. If the blue light and air movement that Monica used as triggers for her pea plants could cause signals to be sent at the same time, perhaps the pea plants could learn to respond with directional growth to the air movement alone.[20] Her results have not been replicated, but the very reason this area of research is so exciting is because of all the unanswered questions it raises. It hints at the kinds of revelations that giving up our zoocentric view of life might yield.

Neurowars

The question of whether plants have "nervous systems" has fuelled a bitter fight for many decades. The debate has had the special kind of understated vehemence that can only occur in academic circles. When I first became embroiled myself, many years ago, the ferocity surprised me. One pivotal experience occurred on a cold, grey November day, when I took a train from Edinburgh to meet a prominent plant physiologist at the University of Glasgow. I arrived about an hour early, and spent it wandering around the campus, past the filigree medieval spires of the School of Philosophy. They were very familiar to me—it was where I had done my own degree many years ago. This building stands just across from the Plant Physiology building, almost next door to it. I remembered one winter during my studies, looking out of the windows of the nearby Research Club, where the graduate students gathered for beer and casual discussions. Thick snow was falling and I marvelled at the sight. Growing up in the Mediterranean, I had not seen snow before. I never imagined that, twenty-five years later, I would be visiting the laboratory just metres away.

Before I took the train to Glasgow, I had thought through carefully what I wanted to talk about. We were going to discuss the recent controversy around plant phytonervous systems. Being a physiologist through and through, my interlocutor was firmly on the side of reducing plants to their molecular and structural properties. He was uncomfortable considering plant behaviour and was certainly not open to the idea of plant *psychology*. I still hoped to find a way to have a productive discussion. Ten years of academic war had been waged on this subject already by different factions who refused to move from their specific intellectual stamping grounds. My main aim was to show that, coming from philosophy and not trained as a plant scientist, my colleagues and I might be able to break the intellectual standoff with a new approach of cross-fertilisation between different specialisms. Perhaps, with open minds, the different sides could be reconciled and could work together to explore the field with new vigour. In a world in which there is ever-increasing pressure for scientists to become more narrowly specialised, more tightly tethered to limited ways of seeing and thinking, I felt particularly strongly the importance of creating these bridges. Being an academic of hybrid pedigree myself, I am deeply aware of the fertile possibilities that can be pre-emptively cut off by overly narrow, laser-like vision.

Unfortunately, we did not manage to build the bridges I had hoped for. Not long after our meeting, a group of leading plant physiologists published a paper attacking our work, without even the veneer of politeness usual in academic publishing. They argued that "dubious ideas about plant consciousness can harm this scientific discipline" and "generate mistaken ideas about the plant sciences in young, aspiring plant biologists." We were not only wrong, they seemed to think, we were dangerous: "serial speculationists" looking to dismantle respectable science from within. They urged funding agencies to refuse us and journals to reject our papers, to keep our "prolific speculating and fantasizing" out of scientific discourse.[21]

There was a deep irony in these plant physiologists' attempts to discredit us. Only a couple of hundred years ago, their own field was seen as faintly ridiculous. It was not until 1856 that the botanist Julius Sachs first focused on plant physiology at the University of Prague, and it took another sixty years before the field was seen as a science and not an idle fancy. But if my colleagues and I thought that public slander and calls for us to be blacklisted by publishers and funding bodies would be as bad as it was going to get, we were wrong. Almost immediately after the first paper criticising us came out, another followed.[22] It referred to "long-distance voltage signals" that influenced plants' "physiological, developmental, and adaptive responses." In short, it described a kind of electrical signalling system in plants, but in coyly physiological terms that shied away from admitting to describing a "nervous-like system," never mind plant behaviour or intelligence. They had shoehorned our ideas into their own reductive structure of plant physiology. And restarted the turf war from decades before.

What's in a name?

The deep consternation felt in certain academic circles about terms such as "plant neurobiology" might be a symptom of an increasingly atomised intellectual world. Meanings are very narrowly defined and ideas are not given licence to interact outside spaces in which they are used in a specialised way. This is worrying, because if you look back at the history of science, some of the most brilliant ideas have emerged from connections *between* ideas, looking at problems with fresh perspectives, and making links between different ways of thinking. Getting overly caught up with labels, in combination with the restrictive effects of historical prejudice, may do more harm than good in the complex project of increasing our understanding of living things. After all, as Nobel laureate Richard Feynman once said, "If we want to solve a problem that we have never solved before, we must leave the door to the unknown ajar."

The concern with semantics runs deep. Many scientists pursuing plant intelligence research with a more open-minded approach are still shy of being associated with "plant neurobiology." Plant ecologist Ariel Novoplansky, some of whose work has been a cornerstone of my own, tells us:

> although this term [plant neurobiology] naturally presents an interesting provocation, it is neither based on solid facts nor does it necessarily help advancing our scientific agenda. On the contrary, the beautiful thing about plants and other "rudimentary" organisms is *that they can do so much without using any neural-based systems.*[23]

Likewise, in an interview for *Scientific American*, plant geneticist Daniel Chamovitz, author of the best-selling *What a Plant Knows*, comments:[*]

> At the risk of offending some of my closest friends, I think the term plant neurobiology is as ridiculous as say, human floral biology. Plants do not have neurons just as humans don't have flowers![24]

---

[*] At the time Dean of the Faculty of Life Sciences at Tel Aviv University and today President of Ben-Gurion University of the Negev.

Even the Society of Plant Neurobiology itself is riddled with contention. The third meeting of the society was held in Štrbské Pleso, in the heart of the Slovakian High Tatras mountains in May 2007. It was the first I'd attended, so I was eager to hear about the emerging field that had begun to consume my imagination since I first read *Communication in Plants: Neuronal Aspects of Plant Life*, a volume edited by František Baluška, Stefano Mancuso and Dieter Volkmann the year before. On the final day of the conference there was a discussion about using the term "Plant Neurobiology." Voices were raised, ranging from staunch proponents to vehement detractors, with little hope of resolution. The chairman of the Society's steering committee, Liz Van Volkenburgh from the University of Washington, Seattle, recalls:*

Plant Neurobiology as a name for our research thrust was provocative and exciting. However, for many good scientific reasons the name provoked divisive controversy. Initially, the organizing committee decided not to change it ... But "Plant Neurobiology" proved to be truly a red herring, and ... the members of the Society agreed ... in 2009 to change the name to match that of its journal, *Plant Signaling and Behavior.*[25]

And that, apparently, was that.[26]

In a commentary for the journal *Nature Plants* in 2018, Chamovitz's comments on plant intelligence reveal the orthodox attitudes of the community. After observing that "plants integrate many external signals to adapt to their environment and increase their fitness," Chamovitz goes on to ask, "Is this a proof of intelligence? It depends on the meaning of the word." But what would be the practical purpose of having a set-in-stone definition

---

* Also founding member and President of the Society for Plant Neurobiology/ Plant Signaling and Behavior.

of plant intelligence? The way other disciplines move forward proves, if anything, that any attempt at anchoring definitions is futile. Biologists, for instance, have coped without reaching a consensus as to what "life" means.[27] Why would "intelligence" be any different? Put it another way: would we be willing to apply the same reasoning to research in *animal* intelligence? Arguing that because we have historically failed to define animal intelligence, we must put research on animal intelligence on hold until a definition is agreed upon, would seem ridiculous. That's not how science works.

While getting caught up with labels is unhelpful, the way we name things can be important. Names carry with them a network of understanding, they frame something in a particular way and guide how we think about it. A word such as "neurobiology" can be used almost tongue-in-cheek, in a very loose, metaphorical sense, to describe the physiology of chemicals and electrical activity in plant cells. Or we can take the idea of plant neurobiology more literally, focusing on the *roles* these processes play in plants as living organisms: electrical signals that integrate information across the plant body. I argue that it would be a mistake to affix the word "neurobiology" too tightly to neurones. This would entail sweeping the similarities in *function* between animal and plant signalling systems under the carpet, because these signals don't occur in the same *structures*. We would inevitably lose the powerful ideas that arise from approaching "plant neurobiology" head on.

One way around the anxieties of talking about the neurobiology of organisms without neurones would be to adapt the meaning of "neurobiology" to be more inclusive. One prominent neuroscientist has done so: Rodolfo Llinás at the NYU School of Medicine. Together with Spanish computer scientist Miguel-Tomé, he argues that while "plant neurobiology" mustn't equate to "animal neurobiology," we could "broaden the definition of a nervous system" to one that "would employ function as a crite-rion."[28] We could define neurological systems by what they *do*, rather than by what cells and tissues carry out these functions. We needn't be trying to make plants into green animals in order to use the same language.

A look at the other essential functions across animals and plants illustrates how this idea works. Animal functions are organised into sets of specialised tissues and organs, or systems. We have a respiratory system for taking in oxygen and expelling carbon dioxide; a digestive system for taking up water and nutri-ents; a circulatory system for carrying important things around the body; and a nervous system for rapid electrical communi-cation. Plants have the same *functions*, but they are organised materially in a different way, distributed more widely. Plants exchange gases through pores in their leaves, take in solar energy and create energy-rich molecules in their leaves, while their vascular tissues serve to transport sugars, water and nutrients throughout their bodies.

The system of contention—the "plant nervous system"—is more elusive. But even the physiologists will admit to the presence of a network of electrical signalling processes throughout the plant as well as many different kinds of sensors for collecting informa-tion from the plant's surroundings and internal environment. And as we have seen, we have clear evidence from recent work that the vascular system is not just a set of delivery tubes for sugars and other molecules.[29] The recent detailed studies of the "neurochem-istry" of plants—on GABA, glutamate and other molecules—are the building blocks of a miraculous whole. The picture emerg-ing, while still partial, is powerful. The animal nervous system integrates the information coming in and triggers coordinated responses to it throughout the organism. And the same seems to be true with the "phytonervous system."[30] The bundles of vascu-lar tissues of xylem and phloem are not neurones, but they have neurone-like features. We have the choice to see that and explore its possibilities to the fullest, or to ignore it and stay within the rigid paradigm of history and continue to argue over semantics.

Maverick thinking

"In science, if you know what you are doing, you are not at the cutting edge. So, if you are at the cutting edge, you don't know what you are doing.'[31]

In a 2009 interview, the Nobel laureate X-ray crystallographer Richard Axel at Columbia University summed up why we need to be bold in our scientific explorations. It is only by thinking outside the box, between the lines and beyond the horizon that we can escape the restrictions of our current thinking and revolutionise our world views in important ways. This is why I at least attempt to be maverick in my thinking. The word itself comes from history: Samuel A. Maverick was a Texas engineer and rancher who did not brand his cattle, unlike most of the ranchers of nineteenth-century Connecticut. The animals came to be called "Mavericks" after their free-thinking owner. I cannot claim to have achieved quite this level of intellectual independence—someone like Bose has far better claim to the title. But it is certainly something to aspire to when exploring new boundaries of scientific knowledge.

Purely reductionist biology is of immense value but has clear limitations as a way of understanding organisms. The Hungarian biochemist and Nobel Prize winner, Albert Szent-Györgyi, illustrated beautifully why this is so. He made the point that if you presented a dynamo to scientists of various disciplines, they would all look at it in different, and limited, ways. A chemist might dissolve it in acid to break it down to its constituent molecules; a molecular biologist would take it apart and describe the components in detail. But, were you to suggest that an "invisible fluid, electricity" flowed in the dynamo, the flow of which ceased when it was dismantled, they might "scold you as a vitalist." Which, Szent-Györgyi emphasised, would be "worse than to be called a communist by an FBI agent"—no insignificant matter at the time when he was writing.[32]

We cannot reduce plants to purely mechanical objects. *Intelligence* is not going to be revealed in physiology, though it is very reassuring to reduce the biological to the purely material. We might better comprehend the complexity of organic processes by likening them to inorganic things, for example. Mechanising the living world to make it seem easier to understand. We could see the eye as a camera, nerves like electrical circuits, the plant's phloem system like a series of pipes. In reality, though, they are none of these things. Such images might be heuristics—making things easier to conceive of makes them easier to deal with and investigate—but they also might trap us in a reductive way of thinking. We have to be wary of ignoring what we cannot easily see.

So if we are to understand fully what makes plants tick, we cannot restrict ourselves to one discipline. Physiology needs psychology. As the prominent American psychologist Edward C. Tolman wrote in the mid twentieth century: "A psychology cannot be explained by a physiology until one has a psychology to explain."[33] The working hypothesis of plant neurobiology is that the integration and transmission of information within the plant involves in part neurone-like processes. Electrical signalling might well play a role in integrating the plant body, linking perception and behaviour, and as such it cannot be ignored when it comes to understanding plants. Physiology and behaviour are inextricable: one affects the other in all organisms, not least in plants. The cells and signals give rise to the behaviour, and the adaptive value of the behaviour is why those cells and signals evolved as they are in the first place.[34] Why should the disciplines that focus on each of these things not work in a complementary way to use the power of both theoretical frameworks to understand both mechanism and effect?[35] We need to put aside the tired drum-beating about whether there is or is not a "plant neurobiology," and approach afresh with open minds and cooperation between fields.

Central to this interdisciplinary mix of ideas must be those from cognitive neuroscience. This field bridges the material and the functional, viewing the important structures and signals in the nervous system from the perspective of working out how they relate to cognitive activity. It is itself interdisciplinary, drawing on many different schools of thought. Even if we framed the physiology of plants in terms of what role they play in interacting with their environments, such as the stress responses that plants mount when threatened by cold or predators, we cannot get a complete picture. Clearly we can't study physiology *without* thinking about ecology, but limiting ourselves to these kinds of links still keeps us in the realm of mere adaptations that we've already discussed. It leaves no space for figuring out where plants are making choices, forcing everything we observe into the box of knee-jerk reflexes. If we can think beyond the concrete, the differences between nerves and phloem, to study intelligent *information processing* by organisms, a whole new world of cognition could open up to us.

Richard Feynman remarked that "science is the belief in the ignorance of experts," meaning that blindly subscribing to the dominant ideas of the past is dangerous. More than any other realm of human thought, science shows us that ideas can and should be continually overturned. We have seen how archaic conceptions of the natural world have blinkered us to understanding very different kinds of organisms—not least plants—in new ways. But the problem may be worsening. In 2014, thirty scientists published an open letter in the *Guardian*. They were deeply worried about the direction that scientific culture was taking: towards narrow focus and mercenary obsession with rapid publishing. There was little space for the emergence of original, lateral brilliance as seen in the big ideas created by independent thinkers such as Feynman, who made twentieth-century science so revolutionary. It might be time to inject a little maverick thinking into science again, and not only in the realm of plant neurobiology.

CHAPTER FIVE

DO PLANTS THINK?

Believing is seeing

Even if you know almost nothing about art, you will have heard of the Mona Lisa. She sits in the Louvre in Paris, smiling enigmatically at visitors who approach to study her in detail. Painted by the polymath Leonardo da Vinci at the turn of the sixteenth century, she is undoubtedly one of the great masterpieces of Renaissance art. But it is not purely her lifelike or aesthetic qualities that have made her so famous and captivating. The "Mona Lisa smile" has intrigued people for centuries, in part because we cannot quite determine what her expression is. Her smile has been one of the most discussed topics in art history: she intimates without revealing. And this mystery speaks to something fundamental in the way we perceive the world around us, which in turn tells us about how other organisms might perceive it, too.

This ambiguity is not only held within the image itself. Whether she is seen as seraphic, happy, wistful, pensive or otherwise has much to do with what those looking at her bring to their interaction with the painting. We see in her what we expect to see. Because, from what we are learning in cognitive studies, how we interpret faces is highly affected by how we feel when we look at them. If you show a completely neutral face to experimental subjects alongside mood-enhancing images, they will be more likely to interpret the face as happy. Exactly the same face, shown alongside mood-dampening images, will more likely be seen as scowling.[1] Our own emotional state determines the emotions we read into others around us, even if those around us are not showing any emotion. Mona Lisa poses a question to the viewer about themselves. Da Vinci's subtle virtuosity left the question carefully unanswered so that, sphinx-like, she might mystify all. The emotions she seems to be expressing are as varied as those experienced by her visitors.

This idea is not limited to ephemeral emotions; it extends to concrete things in the world, too. The way we interpret our surroundings is partly driven by our expectations of what we will encounter. Think of the well-known photograph of a Dalmatian dog barely visible against a spotted background, taken by Ronald C. James.[2] Readers familiar with the image will recall the difficulty in trying to discern something that could be a picture. At first, it surely looked like nothing more than a scattering of black and white strokes, however hard you tried to make sense of it. But once you have identified the Dalmatian dog among the apparently random black smudges, the way the brain interprets the image is entirely altered. It *knows* there is a dog there now, so it seeks out the dots that are part of the dog, needing only the merest suggestion of the canine shape in order to perceive it. The randomness of the picture has been dissolved: your brain clings to the order that has been created.

This kind of experiment feels like a trick of the eyes, a kind of reverse Rorschach test where the hidden image is extended into our perception. But it is quite the opposite. It reveals the inner workings of how we interpret sensory data. The *expectation* of a Dalmatian primes our minds to make one out amongst the previously meaningless dots. It turns out that we don't form images passively from a stream of sensory input, as we might imagine. Perception is not data-driven.[3] Our brains aren't neurological couch potatoes waiting to process information from the world outside. If they were, the knowledge that there might be a dog in the image would not affect its interpretation. Instead, what we perceive is significantly expectation-driven: what we *predict* affects what we will experience. Our brains constantly pre-empt what we will encounter, in a way that shapes the nature of the encounter. This idea might be mildly unsettling. Most people would acknowledge that how we understand abstract things might be affected by our biases and opinions. But the idea that our very perception of the tangible world—even what we *see*— might not equate to what exists, or even equate to the naïve interpretation of the raw input into our senses, seems counterintuitive. Its consequence is that our experiences of the world are far more individual than we realise.

Humans might not be unique in this use of preconception. In 2021, a team at Yale University studied the brainwaves of baby mice, just after they were born—blind and hairless—and shortly after they opened their eyes. While the pups' eyes were still covered by a filmy skin, waves of electrical activity emerged from their retinas. These modelled the kinds of patterns that would occur in older pups as they saw themselves moving through their surroundings. In a sense, we may say that the blind mice *dream* the experience of the world around them, even before they can actually see it. These patterns are replaced with new, more mature circuits when the pups eventually open their eyes. But the blind images allow the young mice to interpret the sudden torrent of visual information, to hit the ground running when they become more independent. When the team blocked the activity

of these retinal cells in blind pups, the mice found it very difficult to interpret moving images and navigate their surroundings
even when they could eventually see. Baby mice, if they have not
been experimentally tampered with, enter a world that they have
already imagined, which was already encoded into their retinas
and minds.[4]

If an internal model is central to how mammals deal with the
world, it may be so for other kinds of organisms, too. We saw
in Chapter Four that you don't need a network of neurones or
a brain to have a "nervous system," so perhaps a neocortex is
not needed for preconception. When a bean sends its fly-casting
tendril around seeking a support, it may be that it is not simply
collecting information and responding. There may be something
altogether more sophisticated going on.

Wild thinkers

More years ago than I care to think, back in the 1990s, I faced
the viva for my Philosophy PhD at Glasgow University. It was
being conducted over an internet video connection, long before
Skype or Zoom existed, which made it even more daunting.
One of the examiners was Andy Clark, who had already made
a name for himself as a visionary philosopher.[5] I got hold of a
preprint copy of a manuscript of his that was in circulation.
Co-authored with fellow rock-star philosopher David Chalmers,
it eventually became the most cited philosophy paper of the
decade.[6] They blew open the insularity of our cranium-bounded
cognition and expanded the mind down to its subconscious
depths and out into the world around us: the objects we interact
with, the other minds that we encounter. Cognition, according
to Clark and Chalmers, included the tools we use to think: pens
and paper, word processors, calculators, art materials—integral

parts of a continuous loop between the internal and external world, between thinking and doing. They described a "self" that was not contained and limited but networked, a composite of neurones and things spanning the inanimate and the animate. The theory caused quite a stir when it was first published. Now, though, that smartphones and other mind-expanding technologies are integral to daily life, Andy's ideas don't seem quite so strange.[7] We export our memories to electronic devices and the internet, and rely on apps for processing functions that our brains used to carry out, such as delegating basic maths to our phone calculators or direction-finding to Google Maps. Our thinking increasingly involves the electrical activity of both neurones and microprocessors.

Little did I know that I would have a chance to share Andy's office two decades later, during my sabbatical in Edinburgh in 2016–17. Working in the same room, I had the luxury of delving into his expansive mind and exploring his ideas on extended cognition. And of learning more about the revolutionary thesis he had developed from these ideas: *predictive processing*. He had turned the conception of the way the brain interacts with the world on its head, arguing that it was not a passive receiver of information, but instead an ever-buzzing "prediction machine" that anticipates incoming experiences. Unlike the baby mice dreaming dreams of the unseen, the brain uses past experiences and sensations to form these predictions, refining them over time. Meeting new incoming information head-on with acquired expectations, he argued, allows the brain to immediately make sense of what it is experiencing. This is called "top-down" processing—in which the brain actively dictates experience—in contrast to the passive "bottom-up" response to sensation.[8]

On my way back to Edinburgh after my meeting at Glasgow University, recounted in the previous chapter, I thought about the conversations I had shared in the office with Andy, and noted how vastly different my interactions with the two people were. I mused on what it took to think truly *wildly*, to imagine beyond the narrow boundaries of discipline. True to his ideas, Andy manifested his creativity not only academically, but through the environment he created. In Andy's office, among a cornucopia of curiosities and idiosyncratic objects, was something which perfectly demonstrated his thesis. It was a "hollow face:" an inverted 3D model of a face, concave instead of convex. To eyes and minds used to seeing faces in the round, the head-on view of the hollow face produces a perfectly inverted perception of the visual distance: the brain predicts a convex face, and so interprets the concave image as such. It was a tangible optical illusion, toying with the brain's expectations. Move to one side, and the face collapses into negative space. I mused on this illusion during the time I worked alongside Andy, enjoying the repeated shock to my visual system as I shifted my visual plane and the protruding face suddenly receded into itself. I began to think wildly myself—to wonder: if our brains make predictions about what they will encounter, in a way that usefully shapes what we experience, might the same not be true of plants? So, why might they not use their expectations to shape their perceptual experiences in a similarly proactive way?

I suggested my ideas to Andy, thinking that they might appeal to his radical imagination. They did, but he did not feel he was quite the right partner for this project. He dealt with the ephemeral products of neuronal wiring in the human brain, and did not wish to venture into the alien world of the phytonervous system. Instead, he introduced me to theoretical neuroscientist Karl Friston at University College London. Just as Andy had become the most cited philosopher in the world, Friston was the most cited neuroscientist—and for good reason. Friston put a mathematical slant on the kinds of biological phenomena that Andy was explaining in terms of predictive processing, related to how organisms use expectation to guide both *how* they perceive and *what* they do.[9] When the input from their perception does not match their expectations, he argued, the brain experiences "surprisal." This is a mathematically defined concept, linked to the *probability* of an event that is abstractly related to being surprised (what you experience when you encounter something unexpected). The brain likes surprisal about as much as it likes bad surprises, though, so it usually attempts to minimise experiencing it.

In Friston's work, the idea that the brain minimises incorrect predictions derives from what he terms the "free energy principle."[10] The brain constantly keeps tabs on the difference between how things seem to be in the world and the model the brain has of it. The bigger the difference, the more "free energy" in the system, which can be minimised by changing what the brain samples from the environment or the model it holds. Put the two together and, by alternating between them, you get a whirring prediction machine which is capable of focusing with laser-like intensity on anything that seems out of kilter with its model of the world, and acting to adjust it. For clarity, we can speak of Friston's approach by dividing it into *active* and *perceptual* inference. Either you make the world similar to the model you have of it (active inference), or you change your model to be more similar to the world (perceptual inference). Most likely, nearly everything you do involves a little bit of both.

Using Friston's model, we could say that cognition is the outcome of the careful combination of these two counter-current flows of information travelling from the brain to the senses, and from the senses to the brain. The resulting experience is a combination of the drives of internal expectation and the correcting influence of any data that is coming in. During our discussions, Friston was open to my idea that these two flows of information might also exist in plants, that plants might both form internal models of the world and use them to guide how they explored it. Much to my delight, we ended up working closely together on a paper published in 2017, "Predicting Green," which explained how these information streams and the way they shaped plant behaviour might make plants into cognitive organisms.[11]

Dealing with surprise

As a thought experiment, let's take three different things found in nature: a tench (a kind of freshwater fish), a snowflake and a daisy. Which is the odd one out? Freshwater fish and daisies are biological systems, whereas the snowflake is not. But what is it exactly that tench and daisies have in common, that snowflakes lack? The answer is homeostasis, meaning "similar state." A tench and a daisy share the physiological capacity to adjust their internal environment, to counteract the destabilising influences of changes in their surroundings and the effects of their own internal workings. They manage to keep their inner environments, if not constant, at least somewhat stable. The relevant aspects of these environments might include body temperature, water levels, pH or any other internal conditions that affect living systems. The snowflake, by comparison, would simply melt were it to encounter temperatures above freezing. More importantly, there would be nothing that the snowflake could do about it.

If you're a tench, the ability to maintain an internal stability will be underpinned by hormonal and neural controls, causing the fish to alter its physiology or behaviour in some way to resist changes. If you're a daisy, hormonal and the "non-neural" activity of the vascular system will play a similar role. Either way, both tench and daisies are in the business of avoiding potentially damaging changes to their insides by keeping themselves within a safe and comfortable zone. Practically speaking, there are no distinctive differences between an animal and a plant, or any other organism, from this perspective. Both daisies and tench use all of the retrospective data they have available to make sense of their personal worlds and the events that occur around them. They use this to create expectations about what their surroundings *will* be like and to act accordingly to enable them to avoid dangerous or excessively stressful conditions. We may think of plants and animals as alike in constructing a model of the world that allows them to make sense of incoming data and guides how they use it.

If a tench is to have a chance to pass down its genes, it had better avoid states that might hold surprises according to its existing expectations. Surprising states for our freshwater fish include those in which it is too dry or too salty to survive. The tench tries to make sense of its local environment and act accordingly in order to minimise its state of surprise. What would our poor freshwater fish do if immersed in salt water? It might swim back into fresh water, re-sampling the environment as it goes, hoping to find that future inputs match its expectations and physical needs. In Friston's language, the fish would be using active inference. Alternatively, our fish may try to change its "model of the world" and adjust those of its internal states that are not in line with the state of the world around it: to somehow resist and withstand the action of excessive salt. The latter, perceptual inference, is probably an impossible feat in such life-threatening circumstances. That would take change on an evolutionary scale that an individual fish would not have the capacity for.

Plants are not unlike freshwater fish when it comes to avoiding salt. High concentrations of salt in the soil stress their roots intensely, interfering with protein synthesis and many other key processes. So plants do whatever they can to prevent salt stress. Most will try to avoid getting into salty situations, seeking out soil patches that chime with their internal models of where they are happy. They try to harmonise their expectations. Roots in particular often exhibit salt-avoiding behaviour, the inverse reaction to that of the foraging roots of pea plants when they encounter nutritious soils. As the delicate root tips venture into unexplored patches of soil, they keep note of the salt gradients they encounter, moving towards decreasing levels of salt that might lead the way to new patches of habitable soil. The gradients are important: roots will be more attracted to a trend of decreasing salt, suggesting that better things lie ahead, than to absolute differences in salt levels, which just mean that this particular patch is all right for now. But if the roots probing in one direction seem to experience only increasingly salty substrate, the state of surprise remains high. The plant concludes that it is barking up the wrong tree, as it were. It gives up the search in that direction in order to seek out pastures less salty through alternative routes.[12]

In contrast, some plants have developed tricks that allow them to withstand salt stress. Over evolutionary time, they have gained the capacity to adjust their internal models of where the acceptable boundaries lie. Not unlike humans, plants display a dizzying array of responses to surprise. A few can actively eject excess salt from their precious growing shoot tips, and also from those of their leaves with prime photosynthetic potential: they don't want to be hampered by their lack of solar panels. Others rely on the retention of water to counterbalance the overload of salt. Mangroves, for instance, can live in very salty conditions for periods of time with no problem at all, because they retain water.

Saltbushes embrace the enemy and store salt in their leaves within special glands, where it crystallises and remains harmlessly.[13] Some plants simply amputate leaves if the stress is too much, like a shocked waiter dropping a tray of glasses. What appear to be physiological responses to stress in fact have psychological underpinnings. The physiological response allows them to survive, but it is triggered by a surprise-inducing mismatch between expectation and experience.

Prediction machines

It is much more likely that surprisal will be minimised if a plant combines Friston's two strategies: perceptual and active inference. Plants constantly perceive and act, alternating between these two modes. They are always adjusting their predictions and modifying their surroundings to make their environment match their expectations. These strategies are not always easy to detect or to distinguish from one another. But Charles Darwin is always there to illuminate. In *The Movements and Habits of Climbing Plants* he observes:

> It has often been vaguely asserted that plants are distinguished from animals by not having the power of movement. It should rather be said that plants acquire and display this power only when it is of some advantage to them; this being of comparatively rare occurrence, as they are affixed to the ground, and food is brought to them by the air and rain.

Here's yet another reason to appreciate the cognitive doings of plants. Salt avoidance suggests the ability to anticipate the environment. Plants investigate their surroundings, gathering salient information. They do this especially when predictions do not match what they encounter. When they do match, and surprise is low, they can relax. If there is a mismatch, plants are spurred on to explore further, to seek out areas which sync with their predictions. Plants do this not only to avoid immediate surprises, but also to reduce the types of surprises they expect in the *future*. They have their own expectations as to what's out there, and they continually keep tabs on any changes to try and stay on top of them, to predict how the world might be down the line.[14]

Therefore plants, like animals, need to use an internal model of their environment before they make any kind of move. We could say they've got to run a sort of *simulation*. What plants perceive depends less on the incoming data itself and more on their expectations of what the world is like: what the sun will do, how salty it is, or how nutrient-dense a host is. Though information flows both outwards and inwards, the dominant direction is from an internal model outwards, resulting in overall perception. Plants lead with their expectations, which will then be checked against incoming sensory information. Like us, they are prediction machines with the ability to self-correct.

Plants need some serious processing equipment to sustain the combination of guesses and corrections flowing in opposite directions. What sort of hardware do they have at their disposal? Plants don't have a brain cortex of processing units arranged in hierarchical layers like we and other mammals are so lucky to have. But plants don't need one: all they need is a form of functional asymmetry between the pathways going in and out, like opposing moving walkways in an airport. As we've seen, electrical communication in plants takes place through the vascular bundles of plants' transport systems, the phytonervous system. And these signals can travel in both directions.

The pathways are also arranged in a sort of hierarchy. If you look at the stem of a papaya plant like the one in O'Keeffe's painting, you can see that it is highly networked, with many connections between the thin vascular tubes. These networks are arranged in layers that operate just like the layers of the mammalian cortex. Our working hypothesis is that predictions flow from the deeper layers outwards, to the superficial sensory ones. And, at the same time, the sensory organs trigger electrical impulses that pass through the outer layers and interact with them. You could say that the vascular system connects plant perception and behaviour in the same way that rapid-fire fibre-optic cables are used in telecomms. But exactly what goes on in these organic cables, the details of *how* they make plants into prediction machines that can be surprised by their environment—that is a question we still have to answer. We can see the physical network but we have yet to understand it fully.

Do plants think?

We have been very slow to begin to appreciate the idea of a plant psychology, though the idea was seeded over a hundred years ago. According to Edinburgh directories from between the 1870s and the 1890s, Shetland's Victorian folk writer Jessie Saxby lived at my house in Edinburgh in around 1883. I discovered this through a letter I received from one Philip Snow. As it turned out, he was a writer preparing a biography of Saxby.[15] He was informing "the current occupant" that Jessie Saxby had lived at my premises, and that he would be delighted to visit and see what the flat was like. Unfortunately, it was my final day as the tenant, as I was packing my car on the very last day of my sabbatical.

When I read the letter, I couldn't have been more perplexed, and yet I was eager to know more about both Jessie Saxby and Philip Snow himself. I opened my laptop and responded immediately to the email address he had provided in his letter. Ensuing email correspondence revealed that Jessie Saxby had been a keen gardener, especially when she left my flat and retired to the Shetland Isles. She would gather wild plants for her garden, and even wrote a number of articles about the flowers of Shetland. Philip sent me a picture of Jessie as an elderly lady, and another of her five sons.

Jessie, Philip further informed me, had an elder brother, Thomas Edmondston, who "was a professor of botany at a very young age." Now I was definitely intrigued. It turned out he was the botanist on board the voyage of the *Herald*, which in the mid-1850s had explored the west coast of South and North America. He wrote a short book afterwards, *Flora of the Shetland Isles*. I rushed to my copy of Darwin's biography by Desmond and Moore, conjuring up images of Darwin and Edmondston meeting in person at some point.

Philip had written:

Thomas only got as far as Sua Bay, Peru, where he was accidentally shot and killed, aged just 20 . . . Another point to note is that although Darwin probably didn't meet young Thomas Edmondston, he was a correspondent with his and Jessie's father, the Shetland naturalist Laurence Edmondston. Darwin sent Edmondston Snr a letter of condolence on hearing of his son's untimely death.

Despite the blow, further correspondence revealed that Jessie's youngest son, Charlie (Argyll) Saxby, had compiled a second edition of the Flora of the Shetland Isles in 1903, and that Charlie had also written an article or book called *Do Plants Think?* I tracked it down online to a sixteen-page reprint in the *Transactions of the Plymouth Institution and Devon and Cornwall Natural History Society (1906–7)*:

Do Plants Think? Some speculations concerning a neurology
and psychology of plants.
  Author: C. F. Argyll Saxby.

Having obtained the full title, Philip finally found a hard
copy on the British Library website which he was able to send
me in late September of 2017. Plant psychology, which seems
so outlandish in the twenty-first century, was already being
considered seriously in 1906 and earlier. Seeing a photograph
of Saxby sitting in front of the very door of my Edinburgh flat
nearly a century before Philip's book was eventually published
highlighted to me just how long it has taken us to reconnect with
plant psychology as a possibility. Argyll's treatment was highly
speculative, but our work here is to turn speculations about plant
psychology into empirically testable scientific hypotheses.

Mindware

I have emphasised the point that physiology only goes so far
in explaining how organisms operate. It needs an overarching
psychology, a higher-level framework that the molecular nuts
and bolts enact. We cannot make predictions about how plants
will behave or what they will do physiologically without viewing them through the same lens as we might view animals in the
cognitive sciences.

  The path to a psychological appreciation of other animal species
has been relatively precipitous over the past four hundred years.
During the 1630s the great French philosopher René Descartes
worked on creating a comprehensive physiological basis for behaviour in humans and other animals. He carried out dissections of
animal parts which he had procured from butchers. At the same
time, he developed detailed physiological theories of how the human

body functioned in a mechanical way, from the workings of the muscles to the operation of the brain. He argued that this mechanical functionality accounted for much of the behaviour of humans and other animals. Most behaviour, according to Descartes, had nothing to do with the mind. The basic mechanisms of avoidance of harm and attraction to beneficial things were enough.

These mechanisms were often based on instincts or a concrete form of "memory." A human would withdraw their hand from a burning heat on instinct. A dog would flinch from music if it was usually played when the animal was being beaten, for example. The mind equated to intellect in Descartes' framework, and, since animals lacked intellect, they essentially operated as complex automatons, their senses equating to the direct effects of sound, light or touch on their brains. There was no need for any cognitive complexity. This physiologically dominated psychology denied any sentience or feelings to animals. It was animal-as-machine. Needless to say, in the Cartesian universe, the idea that *plants* might have any form of sentience was beyond laughable.[16]

About two hundred years after Descartes, the self-proclaimed mechanist Hermann von Helmholtz strayed much further into the realms of psychology with his theories regarding the operation of the senses.[17] He argued that the effects of the senses were indeed material effects on sensory organs and nerves, but that they created an *idea* of things in the outside world. To see or hear or smell was to be conscious, because it entailed a concept of something outside the self. The mind inferred the existence of something in the world from the sensory data filtering in. Likewise, the nineteenth-century French physiologist Claude Bernard, though tethered to the concrete matters of how functions such as respiration, digestion and thermo-regulation occurred, also emphasised the importance of the psychological in understanding the relationship between organism and environment. He argued that the central nervous system in animals connected perception and animal behaviour. Physiology was still foundational for Bernard—he thought that psychological phenomena would eventually be explained physiologically—but this was a move away from the mechanical Cartesian world view. By analogy, we might posit the idea that the plant vascular system also mediates between plant perception and plant behaviour: physiology facilitates psychology.

In the twentieth century, hard physiology had to contend with the rise of psychological focus. Eminent experimental psychologist Donald Broadbent turned the tables on the relationship between physiology and psychology. Previously, psychology had been subordinate to physiology, being seen as the ephemeral extension of the study of functioning of bodily parts. Broadbent argued that psychological theories could be of value on their own, without any physiological underpinnings. Not only that, but physiological understanding might, in fact, be best placed *within* its psychological function. Psychology was gradually becoming the overarching framework within which physiology found its meaning. Likewise, the philosopher Jerry Fodor argued not long after that psychology was a "special science" that could not be reduced to neurophysiology, despite the close connections between the two.[18]

Even an appreciation of psychology might yet be insufficient. Between the material doings of physiology and the descriptive theories of experimental psychology was left the question: what *actually* happens to turn sense data into behaviour? To answer this question, a new kind of thinking is needed. A possible way of exploring the subject might be offered by the computational theories of late twentieth-century scientists such as David Marr, whose work has been so influential in the development of computational neuroscience and artificial intelligence. He argued that descriptions of how neurons are organised and operate in the brain cannot reveal the way that vision or other senses generate perceptions: we need details of the data collection *and* the way it is manipulated. Take, for example, how 2D images collected on the retina are turned into a 3D model of the world in the brain.[19] The physical details are the hardware of the brain's "computer"—they don't explain how the programs work, just as an understanding of a computer chip would not show you how the computer operates. An understanding of the algorithms or the "software" that do the processing involved is crucial. We might have a sense of the ingredients that make up the hardware, but without the instructions for how to combine them into a functioning whole, we can't make a model of the final result.

Turning back to plants, we can conclude that we will not be able to understand them from a purely physiological perspective. Because physiology only offers the hardware, it doesn't show how it operates. Nor can we understand plants from simply observing their behaviour and creating romantic plant psychologies. We need to view them, like animals, as information processors with complex algorithms that turn sensory data into representations of the outside world. And for this, we need to better understand several things. First, we need to know what the parameters of tasks such as support seeking or nutrient hunting are from a *plant's perspective*. What are the inputs to the algorithm that allow it to run? These might not be obvious, or what we assume from our human perspective. Second, we need to begin the complex process of disentangling the set of information-processing steps that happen as the plant integrates data from its *senses* with the *predictions* it has made about the outside world. And third, we need to work out how these are fed back into the behaviour of the plant.

To perceive is to create meaning from sensory experience. This meaning must lead to conclusions about what the world around the perceiver is like and the causes of immediate events. The process allows an organism to shape its behaviour in a useful way: the climbing bean seeking a support must oscillate between its model of its surroundings and the data it collects as it circumnutates to refine the aim of its final grab. We know why it looks for supports, we know some of the physiology of how it does this. But we have yet to work out how these two elements are linked, by what processes *aim* becomes *action*. We have to understand the hardware and the front-end software, the neuroscience and the psychology, but also that which links them. For this, an approach based on information processing such as Marr's might well be invaluable. We suspect that plants think. But only after finding links between physiology and behaviour will we begin to understand *how* they think, to see behind their seraphic, sphinx-like pose.

CHAPTER SIX

ECOLOGICAL COGNITION

Phytosoftware

I'm always very amused by the images that are used to depict plant intelligence. The subject seems to pose a quandary for picture editors. What they come up with says a lot about how plant thinking is understood. In 2005, the journal *Trends in Plant Science* published its March issue on the intelligence of plants, under the heading "Neuronal signalling in plants: Intelligent behaviour?"

On the cover was a cartoon of a game of chess between two sunflowers, one bespectacled individual smugly worsting its desultory opponent. Even a major journal, focused entirely on plants, had to resort to a stereotype of human smartness to portray plant cognition. A decade later, things had not changed much. In December 2014, *New Scientist* urged its readers to rethink plant intelligence under the heading "Smarty Plants." This time, they illustrated the idea with a brain-shaped potted plant engaged with Rodin's famous statue *The Thinker*, a classical male figure stooped over, resting his chin pensively on his hand. The subheading read: "They think. They react. They remember." As ridiculous as they are, these clichés tell us something about how narrow the dominant view of intelligence is. As I was first preparing material for this book several years ago, my son kindly designed a book cover for me, bearing my original idea for the title. Of course, in order to represent clever plants, he reached for his own image of where intelligence was best demonstrated: the classroom.

PLANT COGNITION,
THE NEXT REVOLUTION
Autor: Paco Calvo

We have always relied on metaphors in order to understand thought. It's too ephemeral a thing to be able to conceive of it directly, we need a way to make it concrete enough to think about. Each era has had its own metaphor to represent intelligence, often using the dominant technology of the day: from pumps and water clocks to clockwork and telephone networks. We've long done so to make sense of human and animal intelligence, and now we do so to try to come to terms with the cognitive life of plants. But it's not a one-way street. The image is a tool for thinking but it inevitably shapes the ideas it engenders. In fact, we've taken it a step further, quite literally using computers to simulate intelligence. As a Fulbright visiting scholar in the late 1990s at the University of California, San Diego, I was fortunate to witness at first hand the surge of artificial neural networks. The decade from 1990 to 2000 was dubbed the "Decade of the Brain."* Artificial neural network modellers teamed up with neuroscientists to model cognition by adjusting the "synapses" within an artificial neural network. Inspired by the *functioning* of the human brain, abstract mathematical units would stand for biological neurones, and numerical connection weights for synapses.[1]

The preferred metaphor of plant scientists, unsurprisingly, is the digital technology that rules the infrastructure of the human world. So plant intelligence becomes a story of computation. It suggests that, if plants are intelligent, it is because they process information. In the very same way you can play chess with a computer that follows software routines, "smarty plants" are able to interact with Rodin's thinker because they *compute*. Nature must have installed some kind of "software" in plants which allows them to act like green computers, sifting data from their surroundings and processing it to produce behavioural outputs. Of course, the journal editors used pictures to give a light-hearted twist to the complexities of internal signalling systems. But the computer metaphor is also taken quite literally. It implies that if you understand the rules written in software, you understand cognition. This is the essence of David Marr's computational theory of mind.

It's a useful metaphor on some levels: I have been using the terminology of "hardware" and "software" here myself. To take a closer look at what this kind of mechanism boils down to, what *computation* means, is most easily done by going back to the ancestors of today's computers, such as the "Analytical Engine" devised by Charles Babbage, a contemporary of Darwin. This was a theoretical invention—the engine was never actually built. Babbage found inspiration from a weaving loom designed by

---

* The 1990s were designated the Decade of the Brain (DoB) by the US Congress. Originally sponsored by Congressman Silvio Conti (R-Mass.) on the recommendation of leaders of the neuroscience community, the proclamation was signed by President George Bush in July 1990.

Joseph M. Jacquard for the textile industry. Jacquard's invention consisted of an ordinary loom with a set of cards attached to it. The idea was to automate weaving patterns into cloth. Perforations in the cards corresponded to the desired patterns. By arranging the cards in a sequence, and feeding them to the loom, the rows of the design could be woven one after the other.

The perforations in the cards constituted the software—or set of rules—that ran on the loom. It removed the need for the mind of a weaver to pay constant attention. Babbage imagined building a calculating machine in a similar manner, a complex steam-powered assemblage of cogwheels and shafts using punched cards.[2]

The Engine's power rested in the punched cards, its software. "Number" cards held values, "variable" cards placed values in columns, and "operation" cards chose actions—say, division or multiplication. In addition, the engine had a store to hold the numbers and a mill to process them—which would be the memory and the central processing unit of a modern computer. In essence, the analytical engine was a general-purpose computer with mechanical parts instead of electronic ones, but it reveals the principles underlying the powerful technologies that came after it. Change the pattern of holes in the cards, have the machine read different batches, and you are effectively running a different program onto the engine. Ada Lovelace—considered the first programmer in history and daughter of the Romantic poet Lord Byron—remarked in her notes on Babbage's project: "we may say most aptly that the Analytical Engine weaves algebraical patterns just as the Jacquard-loom weaves flowers and leaves."[3]

The computer metaphor may have cast a shadow that's far too long. It implies that thought is similar to the type of regimented data-crunching that takes place in a chess game. The implication is that if plants cognise, they must be following some set of rigid instruction pathways that allow them to sense and react to their environment. But chess is a *formal* game made up of a set of simple rules. Computers excel at manipulating large pools of data, under such explicit rules. In other words, it's not the material of the pieces that matters—it is the rules being followed. To play a game of chess, you don't even need a board, you can just use a screen or even a series of numbers and letters. Rules do not matter for plants, though. A machine might "weave flowers and leaves," but it cannot recreate the living organs.

Today's supercomputers dwarf Babbage's Analytical Engine, but we shouldn't be too impressed. The information-processing arms race boils down to how many instructions per second you are able to handle. This may have very little to do with biological intelligence. There is something that makes us feel uneasy about reducing the incredible complexity of our mental life to a piece of software. Instinctually, we feel that we are more than automatons running complicated programs. Perhaps this is one reason why artificial intelligence (AI) seems eerie and unnerving: it enables machines to *appear* to function like us, but in a way that is underpinned by mere computation. They become inflexible doppelgängers with capabilities that both fall far short of and far exceed our own.

Mind is matter

If you compare a game of chess with a game of pool, you might notice they could hardly be more different in terms of action. One involves a set of rule-bound calculations, the other interactions between cognition and physical action. Pool players can't just strategise; they must enact their thoughts on the pool balls with a cue. Thought must extend into the material realm, and in real time. It is possible that we will never be able to describe the *thinking* of either plants or animals with formal rules like a game of chess. We can't just look at the software, or the physiological hardware. We must take into account that plants and other organisms are physical beings that exist in a network of tangible interactions. Perhaps plant thought is far more like a game of pool. It has to be understood through physical changes in its ecological surroundings.

What's the difference between describing behaviour *with* rules and seeing behaviour as the *result* of rules? Think of bees and the hexagonal shape of their honeycombs. How do bees create hexagonal structures out of spherical wax cells? What rules do they follow? It is tempting to credit the bees themselves. Darwin did suspect that bees manipulated cells that were initially spherical, but he never observed the process of transformation. In his desire to see nature's entirety through a unified lens, he posited the idea that bees made hexagons as the result of natural selection: that is, that the honeybees that built their hives out of hexagonal cells were most likely to survive and reproduce. However, bees do not intend to build the hexagonal structures of their natural honeybee combs. Bees do not refer to an inbuilt rulebook. We now know that hexagonal cells are the result of *physics*, not of evolutionary biology. Bees pile up spherical chambers of wax, and as cells pile up, they compress. The surface tension where the cell walls meet forms hexagons spontaneously. No rules for making hexagons are being followed.[4]

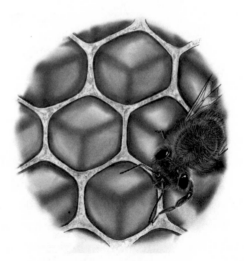

Complex hexagonal shapes, which look like they might be the result of a computer-driven 3D printer, can also be the product of bees' urge to pile wax spherically and the laws of physics. So perhaps focusing on "software" as we currently do is insufficient when we seek to understand cognition. As we have seen, the physical world is almost irrelevant to the activity of computers. A computer can be a tiny microchip or a supercomputer that fills a whole room, a smartphone or an AI robot. In contrast, living beings—and their minds—are deeply connected to their physical forms and the world around them. A parakeet's brain would not operate the same way if inserted into a mouse, nor would a beetle's consciousness be transferable to a petunia.

The American computational psychologist and winner of the Nobel Prize in Economics, Herbert Simon, summed up this dependency perfectly with an influential parable about ants running about on a beach. Imagine observing their behaviour as they trundle haphazardly across the sand. In isolation, their progress looks erratic and complex, the ants wending this way and that with an apparently random walk guided by rules that seem impossible to deduce. But consider the terrain the ants are walking over, and you realise that the ants are doing nothing more sophisticated than avoiding obstacles. The apparent complexity of their paths is due to the nature of the environment, which dictates their behaviour. The tiny ants are simply seeking a way through what is to them a barely traversable terrain of sand dunes.[5]

This is not to say that all behaviour is simple. Rather, the ants on the beach show that behaviour must be understood within the world in which it is enacted. Animal-centric, material physiology shouldn't define what cognition is. And, at the same time, the mind is not immaterial. In their "Extended Mind" thesis, Andy Clark and Dave Chalmers generated a picture of cognition that expanded out into the world, beyond the confines of the cranium, incorporating the tools and things we used to think with. Plants don't, of course, have heads, but they do expand into their worlds with roots and shoots, tendrils and suckers. They grow into their environment, collectively creating the green infrastructure of the ecosystem, mingling with bacteria and fungi under the ground, battling predators at the margins of their leaves and along the lengths of their stems, sending packages of sex cells great distances on the bodies of animals after fleeting dalliances at their flowers. Perhaps their "minds" extend into the world in just the same way as Clark imagined ours do into our smartphones and pencils and Lego blocks. The approach of *ecological* psychology takes this a step further, seeing the physical nature of living things and the physical nature of their environments as *integral* to their thinking.[6] And not least for plants.

At MINT Lab, we don't focus on what the rules are for controlling plant behaviour; instead, we look at how the relationships between plants and their surroundings affect what plants do. As psychologist William Mace put it: "ask not what's inside your head, but what your head's inside of."[7]

Bridging distance

The Lumière brothers made one of their first films, *L'arrivée d'un train à La Ciotat* (The Arrival of a Train) in 1895. The silent documentary showed the entrance of a train to La Ciotat station, the passengers disembarking with staccato steps. Some reports attested that when it was first shown in cinemas in 1896 to audiences almost entirely unused to viewing moving pictures, the sight of the train rushing towards the audience on screen caused several people to scream and run for their lives to the back of the auditorium in panic. This may well be an urban legend, but it is a very believable one. We live today in a world where so much of our consciousness interfaces with screens, inured to their constantly moving images, that we can't imagine being scared by an approaching train as if it were real. But in the eyes of nineteenth-century cinemagoers, not yet jaded by over-exposure to film, the sight of a train rapidly hurtling closer, even in monochrome, could have triggered the most primitive of impulses.

Why exactly would this early film audience have been so terrified by a two-dimensional image? Surely, they could see that the screen itself did not get any closer. Nothing was *actually* moving towards them. It comes down to how we assess what is going on around us. One of the problems that organisms face as they interact with their surroundings is judging distance. At a basic level, their physical forms need to make contact with and avoid objects in the world around them. They need to know what the relationship is between their own forms and other objects in dynamic, sometimes rapidly changing situations. It might seem to be a different problem for animals and plants. Compare a tamarin monkey leaping between branches accurately to avoid falling to the forest floor and a circling vine lassoing a pole to climb up from its rooted position. Surely they can't be facing the same challenge or using similar solutions? But when we examine what's really going on, it turns out that they may well be.

In the science fiction novel *The Black Cloud* (1957) by British astrophysicist Sir Fred Hoyle, a gigantic cloud of black gas appears ominously near Earth. Photographic plates taken in the fictional observatory at regular time intervals reveal that the apparent size of the cloud keeps growing. The cloud seems to be encroaching on Earth in a worrying manner. A group of astronomers make the educated guess that it will *eventually* collide with planet Earth. If only they could estimate the distance and speed of the approaching cloud, they would be able to calculate the time before the arrival of cosmic Armageddon, and take appropriate action, whatever that might be. One astronomer reasons that they could calculate the cloud's speed using the light emission spectrum from the stars being blacked out as the cloud grows. But that proves unnecessary because there is a far simpler solution. In order to tell how much time they have left, there is no need to estimate either the speed the cloud is travelling at, or its current distance away. They can simply look at how the apparent size of the cloud is increasing.

Let's consider how this works with a simple, Earth-bound analogy. Take a look at the two bricks in Figure 4. Since they are different distances away from you, their projected sizes on your retina, for which the camera is standing in, differ. The further away it is, the smaller a brick appears. Yet you don't take the one at the back to *be* smaller. You *perceive* them both as the same size. Psychologists of perception call this the "size constancy" problem.

Figure 4: Example of size constancy: a texture gradient formed by cobblestones with two bricks. The space occupied by the bricks is the same with respect to the area being covered.

How do you explain size constancy *ecologically*? Your optical system and brain don't go through the rigmarole of computing distances and inferring brick sizes. They have no way of collecting that information, for starters—we are not machines equipped with speedometers and tape measures. Another type of information which is just as useful is directly available at the retina itself. Notice that the percentage of cobblestone occupied by each brick is the same: both bricks cover about a third of the cobblestone they stand on. Regardless of the different distances and the projected sizes on the retina, it is the *ratio* of brick width to cobblestone surface that provides the clue which allows your brain to understand that the bricks are actually the same size. It's not absolute length and distance that we build our perception from. Instead, it is the relation *between* objects and their environment that we use.

The imaginations of science fiction writers can sometimes throw up scientific insights and predictions. Hoyle was a bit of an *ecological* astronomer. His characters used the sensed ratios to work out the cloud's movement, and how long it would take to arrive. Picture the cloud as a basketball being thrown at you (Figure 5). The distance between the ball and your eye is decreasing rapidly. Let's assume that at time t the basketball is at $z(t)$ distance, and that it approaches at a constant speed. The image the ball will project on your retina will have a size, $r(t)$, that is proportional to the real size of the ball. As the basketball approaches you, its image will grow at $v(t)$.[8]

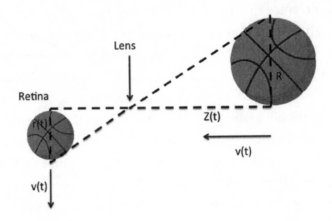

Figure 5: Geometry of an approaching basketball.

Now, if you take the ratio of the size of the ball image projected on your retina to the rate of change of its relative size (its rate of expansion as it approaches your face), it turns out you have all you need. The two triangles in Figure 7 are similar: the ratio of *r(t)* to *v(t)* approximates the ratio of *z(t)* to *v*. Let's call that ratio τ (the Greek letter *tau*), a label coined in the 1970s by Dave Lee, an emeritus professor at the University of Edinburgh and long-term collaborator at MINT Lab. τ is a relative measure of how a gap between any object and an observer is changing. Think of τ as proportionate to the time-to-contact at the current speed of travel.*

If you are doing a physics problem on paper, when a ball is moving at a constant speed, the time it takes to close a gap is ordinarily calculated by dividing distance by velocity. If you are an organism sensing its environment, simply seeing that ratio using the changing image on your retina reveals the time left before the basketball hits your face.

Switch Hoyle's black cloud approaching the Earth back in for

---

* For the mathematically minded, τ is the inverse of the rate of expansion of the image of an object as projected on the retina. The equation would be: $τ = r(t)/v(t)$.

the ball, and the same relationship holds. For Hoyle's threatening cloud, provided that the rate at which the cloud is approaching planet Earth remains constant, τ will tell the time-to-contact. And τ is revealed by the series of observatory plates taken at regular time intervals, acting like a giant retina—you only need to know how much larger the size of the cloud is on one plate than it was on the last one. The rate of dilation of the image gives it away. The flustered astronomers in Hoyle's science fiction calculate that the cloud's image has increased by 5 per cent between the first plate and the second plate taken a month later. They predict that the cloud will reach Earth in twenty months. I won't spoil the book for you by telling you what exactly they do about it.

The Black Cloud is science fiction, but ecological reliance on ratios is common in the real-world doings of living things. Back in 2017 in Edinburgh, as my sabbatical was coming to an end, Dave Lee and I had decided to change our customary weekly beer together for a stroll at the Bass Rock on the Firth of Forth. It was the same coastline where, as a student, Darwin would wade through tidal pools looking for sponges, sea pens and other marine treasures. We sat by the granite cliff on the mainland for hours, blustered by the wind, watching a flock of gannets plummeting one after another into the sea. They seemed to fold their wings back with rapid, mechanical precision, turning from birds into pale arrow streaks in the instant that they entered the water. I imagined them plunging down into the schools of fish below like harpoons, trailing slipstreams of bubbles as they grabbed a fish and sculled to the surface. Dave told me that this was only about the fifth time, in the four decades he had been visiting the spot regularly, that the birds had put on such a show. It felt like they were giving me a spectacular send-off.

Bass Rock was not a casual choice for our farewell stroll. Dave had visited the place regularly over his many years in Edinburgh at the forefront of research into movement in human and non-human animals. By now, he well knew what the gannets were up to, how they managed to pierce the water so perfectly. In a paper published in *Nature* in 1981, Dave and former student Paul Reddish analysed films of diving gannets.[9] They wanted to know: how can gannets tell the precise moment in which to fold their wings and avoid damage or even breaking their necks before entering the water? Their eyes are located beside their bill. Having binocular vision, they can gauge distance. However, Dave saw that Hoyle's science fiction insight might apply to gannets. In the same way that astronomers did not need to know the size of the black cloud, gannets might not need to care about distance and velocity. As it turned out, he was right: gannets are sensitive to τ. The birds use changes in the size of images on their retinas to gauge time, folding their wings at the right time-to-contact. They never need to know what their absolute velocity or height is. τ conveys the *relative* changes, which is all they need to know their time-to-contact with the sea's surface.

Seeing is knowing

I had talked to Dave at length about my studies of the climbing bean and its circling searches for supports. He suspected he knew what the climbing bean was doing as it reached for a nearby pole—it was something not very far removed from the precipitous diving of gannets. That is just one example of how animals use their movement to perceive their environments. Bees approaching the landing strips on flowers, a grey squirrel doing acrobatics to get to a bird feeder, pigeons bobbing their heads as they move—they're all using motion to trigger image changes on their retinas. This gives them a handle on τ, and allows them to intuit time-to-contact. Animals are extremely sensitive to this measurement. It's a vital piece of information.[10]

Perhaps, we wondered, plants work in the same way, perceiving the relative changes in things, like the roots seeking out less salty patches of soil. There is no reason why ecological information would be exclusively available to animals. They generate visual information on their retinas, which are an animal trait. But if plants collected similar information, could they use it? We know that plants are not still life. They move continuously, if only by changing their shape rather than walking. If you imagine a bean fly-fishing for a support, whipping back and forth, getting closer and closer with each cast, you might be able to picture why having some sense of time-to-contact might be important for a plant. And how it is able to collect this information: it's moving through its environment, changing the relative position of the pole to its stem.

Darwin, once again, foresaw that plant behaviour boiled down to tuning into relational differences. In the concluding remarks to *The Power of Movement in Plants*, he observed:

We found that if seedlings, kept in a dark place, were laterally illuminated by a small wax taper for only two or three minutes at intervals of about three-quarters of an hour, they all became bowed to the point where the taper had been held ... Wiesner ... has shown that the same degree of curvature in a plant may be induced in the course of an hour by several interrupted illuminations lasting altogether for 20 m., as by a continuous illumination of 60 m. We believe that this case, as well as our own, may be explained by the excitement from light being due *not so much to its actual amount, as to the difference in amount from that previously received*; and in our case there were repeated alternations from complete darkness to light. (emphasis added)

We can now think about what it actually means when plants "anticipate" something. Does plant planning rely upon the ability to create an internal *model* of the environment? Not necessarily. Astronomers and gannets can predict the future without computation, telling time-to-contact from direct observation. By the same token, plants exploit ecological information that betrays what is going to happen. If certain conditions remain the same, plants, like gannets, can guess the future with great accuracy. The information is there, available in the patterns of change they detect. Animals and plants sense the opportunities for actions that their surroundings offer. For animals, that might be anything from landing sites to prey, while a climbing bean locates a structure that could be climbed. There is nothing magical or computational about anticipation: the interactions between the physical environment and the senses, and sensitivity to relative changes, give all the necessary information.[11]

Piloting vines

This story goes back to the American psychologist who fathered ecological psychology in the 1940s, before Hoyle dreamed up his black cloud. J. J. Gibson realised that there was another way of thinking about the information collected by the retina: it might not judge when an organism will impact with a particular object, but about how the eye is moving through the world. Dave Lee visited Gibson as an early post-doctoral researcher in the 1960s, and owed the seeds of his *tau* theory to him. Gibson's pioneering thinking was spurred on by his desire to solve a problem which had plagued the US Air Force for a long time: how to train pilots to *see* in a way which would allow them to navigate across deserts, oceans and other landscapes, to execute tricky manoeuvres such as spins and landings. Most people are used to judging space on the scale of rooms and streets, and they become deeply confused by the vast spaces around which pilots have to navigate. You have probably experienced this yourself, looking across a landscape and battling the sense that you could almost reach out and touch parts of it, even though you know they might be a great many miles away. Gibson developed a program that would help train prospective pilots to fight their natural instincts for judging distance and to pay attention to different elements of their visual experience which would allow them to navigate the airspace over unfamiliar terrain.

When objects are very far away, τ is not going to be of much help. The relative changes in size on the retina are so small that they cannot give the accurate time-to-contact intuitions that we usually rely on. But Gibson's method relied on something equally instinctual, which you will have experienced yourself. Think back to the last time you were in a car: the world outside the car seems to expand around you as you move forward on the road. But this happens unevenly; objects closer by move towards you and past you faster than objects further away. Road signs hurtle forwards and whip past while distant landmarks sail very sedately along, until finally they are behind you. The horizon barely changes at all—it might to all intents and purposes be static and flat because it is so far away. Gibson called this relationship *retinal motion perspective*. It describes the central way in which we collect information about how far away things are. As we move through the world, we continually compare how fast things move in relation to other objects in our visual field. Closer objects change in position more rapidly relative to us than distant objects do, so they flow more quickly across the visual field.[12]

Gibson used this idea to train pilots to make judgements about aerial space, in an Aviation Psychology Program report for the US Army Air Forces that was eventually declassified. But he was well aware that it would be valuable in other fields too.[13] The way he described the relationship between pilots and their environments applied directly to animals moving through the world: dragonflies sailing along, observing retinal motion perspective through the thousands of tiny lenses in their compound eyes, or gazelles running across the savannah. Their legs or wings or any other locomotive organs act as part of their perceptual system. What they perceive, the things they collect the most data on and see most clearly, are determined by the decisions they make about where they *move*. Likewise, a plant will grow in a direction that seems of interest, and so learn more about that particular part of the world, by observing the relative changes that happen as it moves into it.

Being a plant might well have some fundamental continuities with being an animal. They both move through their environments, collecting information as they go; they use the relative changes they detect to *sense* future changes. It's not difficult to picture a honeybee piloting through a flower-filled aerospace, dwarfed by the plants around it like one of Gibson's Air Force trainees flying over an open landscape. It accelerates forwards and backwards as it visits each flower, making fine spatial judgements as it aims for the nectaries, using the petal shapes and fine scattering of pigment dots that act as landing guides to help navigate. When fully loaded up with saddlebags of pollen and a stomach full of nectar, it zooms back to the nest, taking in the rate of retinal or optic flow of its surroundings to guide its judgement of distance. We find it harder to see plants as pilots navigating space as they grow slowly into it, or move their limbs around in the air, but we hope to apply to the navigation abilities of plants the principles of optic flow that explain how insects navigate. They might not always move as quickly as the bee, but they still might be sensitive to the approach of objects or the rate at which they move past. Our climbing bean is doing nothing if not navigating unknown regions when it circumnutates to find a support, using its own relative movement to gauge where things are around it. The final grab as it lassos its pole is not unlike the bees nearby making assured and rapid entrance into the cavern of a flower's petals.[14]

PART III

BEARING FRUIT

"I *wish* you could talk!"
"We *can* talk," said the Tiger-lily: "when there's anybody worth talking to."

Lewis *Carroll*, Through the Looking Glass

CHAPTER SEVEN

WHAT IS IT LIKE TO
BE A PLANT?

In 1974, the philosopher Thomas Nagel asked, "What is it like to be a bat?"[1] This strange question stimulated decades of tongue-in-cheek additions to the long-standing conversations about the nature of consciousness. Why would we wish to know about the interior worlds of bats? But he had good reason to consider what it is like to be a bat. They are, like us, mammals—and not so distantly related to humans that the idea of them having rich subjective experiences would be impossible to imagine. But on the other hand, their mode of being is also radically different from ours. As Nagel put it: "Anyone who has spent some time in an enclosed space with an excited bat knows what it is to encounter a fundamentally alien form of life."

Nagel argued that if there was something that it was like to *be* a creature, then that creature must have some form of consciousness. Subjective experience equated to consciousness at some level. Rats, whales, antelope, all have specific kinds of internal experiences, tied to their individual ways of perceiving and being in the world. But the question also poses quite a challenge: how can we, with our bipedal, tactile, vision-oriented existence, know what it is like to be a hyperactive, airborne insectivore with webbed digits that "sees" through sonar? We might accept that there is a subjective "bat experience," but the bridge to understanding it is long and tenuous. Picturing what it is like to be a plant—that is even more of an extreme leap of imagination. So much so that many argue that plants have no subjective experience at all.

How can we imagine the experiences of a radically different life form, when its way of existing is so far removed from ours? The problem seems greater when you consider organisms much further away in evolutionary distance. Nagel's bat was not so "far down the phylogenetic tree" that the task was rendered impossible. There have been plenty of attempts to use sophisticated gadgets to simulate what life is like from a "bird's eye view" or the perspective of a fish. We can gather footage and sound from realms very distant from our own and render them into neat experiences that can be consumed from the comfort of the sofa in nature documentaries, or by wandering around museum galleries. National Geographic's Crittercams, for example, use robust recording devices attached unobtrusively to choice subjects—the fin of a shark, the back of a penguin, or a turtle's shell. They provide us with the sights and sounds of these animals' daily experiences.[2]

Technology has made some assaults on the private lives of plants. In addition to time-lapse photography condensing plant growth into visible movement, the artist Alex Metcalf designed a hyper-sensitive microphone to record the noises of tree transpiration, transducing the imperceptible into the audible realm for human ears.[3] These technologies allow us to peer through the keyhole of other species' experiences. But they only put *human* perception in their place: images that fit on a widescreen TV, the wavelengths of light that can be detected by the human eye or sound frequencies the human ear is responsive to. They don't show us what the organisms are perceiving and feeling. How do we make this transition, from a plant-like *perspective* to plant *experience*? The answer is not clear yet, but now, having carefully pieced together what we know and don't know about plant cognition, we are in a position to start considering what it might be like to be a plant.

Shifting perceptions

Philosopher of mind Frank Jackson conjured up an alternative thought experiment to Nagel's bat in 1982, one which cut to the heart of the problem of neuroscience itself. He imagined a neuroscientist, Mary, who knew everything there was to know about colour. Her only limitation was that she had been raised in a black and white room, watched only black and white television, and read black and white books. She knew all about the science of colour, but had never *experienced* it. Jackson argued that Mary, therefore, had a crucial void in her understanding of colour, because there were aspects of understanding colour that could not be described by the discipline she was immersed in. How could she possibly really *know* what colour was? Jackson argued that she couldn't. He insisted that the imaginative capabilities ended abruptly at the limits of her subjective experience, and no amount of technical understanding could make up for that.[4]

Extrapolating from Mary, perhaps even knowing all we know about plant neurobiology, we too can never really understand what being a plant is like. But not everyone agrees with this. My supervisor during my Fulbright scholarship in the late 1990s at the University of California, San Diego, was Paul Churchland, an eminent philosopher of science. He suggested that Mary, being a creature with an imagination as well as an intellectual neuroscientific grasp of colour, *could* put herself in the shoes of the colour-perceiving people around her.[5] Our imaginations, Paul argues, are capable of dramatic leaps into other worlds, especially when coupled with a plentiful scientific understanding.

In one of his lectures, Paul gave an example which illustrated his belief. He told us about a radical shift that he and the other residents of Canada had undergone in the 1970s, when the entire country switched from using the Fahrenheit to the Celsius scale. The Canadians all had to recalibrate their internal sense of temperature, learning a new way to quantify their personal experiences. A hot summer day was no longer 100 degrees, it was only 40; a frosty cold snap was now a frigid minus 10, not 14 degrees. It took some time but they managed it, slowly adjusting to a new set of instinctual numbers. The US, on the other hand, remained stuck in Fahrenheit.

Paul made the point that, as much as it might have caused great consternation, this was not really such a big shift; it merely entailed learning to attach new numbers to the experience of warmth or cold. What *would* have been dramatic would have been learning to calibrate this experience with the average kinetic energy of the particles in the air, or even their mean velocity, which is really what temperature is. This change would have allowed them to tap into the framework with which we describe particles in physics. It would also offer a greater understanding of why the atmosphere behaves as it does: the weather we constantly complain about is the result of the minute activities of the atoms in the air, after all. We could take this further, Paul suggested, and replace the musician's scale of pitch with one of the wavelengths of sound waves, or replace the language of colour with a vocabulary using the wavelengths of the electromagnetic waves that make up colours.[6]

Creating an instinctual relationship between our experience

of warmth and particle speed, or appreciation of colour and light wavelength might seem like a dark art. But altering the way we structure experiences is entirely possible, as the residents of Canada demonstrated. Particles of gas moving at greater or lesser velocities are going to have very different effects on the sensory organs of the creatures wandering through them. The qualities of different wavelengths of sound or light have direct, observable effects on our sensory systems. All it would require would be some training to adopt new frameworks to conceive of these experiences.

Similarly, connecting what we know about plant biology and what it is like to *be* a plant might seem difficult. Churchland's suggestion for understanding the minds of other animals is that we can and must use our minds to shift the perceptual framework, as we might learn to change how we think about temperature or sound. His ideas might also be applied to plant experience. We cannot possibly comprehend what it is like to be a plant, unless we are willing to give up anthropomorphising, to break out of what it is like to be a human and imaginatively explore other ways of being and understanding the world. A complete understanding of the concrete neurobiology will be key, and this project is underway. We need to think outside our own narrow intuitions of what is relevant in our environments and try to imagine it from other *kinds* of consciousness: like bats living in a dark world illuminated by sonar, or plants drawn to the nourishment of sunlight or the mineral richness of the soil.

Learning from cephalopods

Before we make the leap across a vast evolutionary space, let's begin by exploring how we might probe the mind of a creature that at least possesses a head—which is no easy task. Nagel might have thought that a bat was a "fundamentally alien form of life." But if bats are like aliens from another planet, some animals hail from entirely different galaxies. Yet they can show incontrovertible signs of having complex interior lives. Octopuses are a very special kind of mollusc that live without the protection of shells. Instead, they have developed something akin to the vertebrate head. It is very fitting that they are part of a group called the cephalopods: "cephalopod" comes from the Greek for "head foot," and octopuses do at first glance appear to be large heads connected directly to legs. Octopuses are paradoxical: existing on a very short timeframe, living only a year or two at most, they have the kind of intelligence that we'd expect of animals that lived a great deal longer. Their brains are much larger than those of mice, with about forty lobes, including one that seems to function similarly to the frontal lobe in mammal brains. They solve complex problems with causal reasoning and use objects like tools. They ad-lib ways to hunt or avoid predators, and even seem to interact with humans as if they are aware of others' mental worlds.[7]

At the same time, octopuses are incredibly different from us. As the philosopher Peter Godfrey-Smith puts it in his book, *Other Minds*, "meeting an octopus is probably as close as we will come to meeting an intelligent alien."[8] A striking difference between us is the way that their consciousness seems to be diffused through their body. Each of their eight arms can operate independently of the central brain, having its own ganglia and neuronal networks. Their arms can be guided by the eyes to do tasks, but the processing that happens in the limbs seems decoupled from the processing in the brain or even that in the other arms. They have, in some respects, multiple brains. How can we conceive of what it is like to be an octopus—a creature with more than half of its cognition in its limbs and potentially plural consciousnesses?

The filmmaker Craig Foster got as close as anyone might have to accessing the interior world of the octopus and produced a film of his experience in 2020. *My Octopus Teacher* documented Craig's year visiting a common octopus in the kelp forests of False Bay near Cape Town in South Africa. He went far beyond the Crittercams and recording devices, fully entering the world of the octopus with barely more than his flippers and a snorkel, making contact with her almost every day of her brief lifespan. If anything might allow us to shift our frame of reference as Churchland suggested, to enter the mental states of a wildly different creature, then developing an intimate relationship with them is possibly one of the most powerful. He described "in my mind thinking like an octopus," an interaction that went so far beyond the observation of her activities and behaviours that it became an ongoing shared experience between them. This meeting of minds happened in her watery environment rather than his terrestrial one: Foster could only learn from the octopus if he truly entered her world.

Craig Foster managed to get as close as a human might do to understanding the experience of an intelligent marine beast with a decentralised form of intelligence. And perhaps we can learn something from this communing between human and cephalopod. In many ways, the plural brain and hydrostatic form of the octopus is not so different from the fluid body plan of a vine, through which its particular consciousness diffuses. Just as the octopus can carry out many functions recognisable as "conscious" with a vastly different type of nervous system from a mammal, plants too might have converged on similar cognitive abilities with a "phytonervous" system of the sort spelt out in Chapter Four. The ganglionated arms of the octopus are not unlike the tendrils of a vine probing intently into space. To shift our perception, our frameworks for understanding conscious experiences of different kinds, we must also step into the worlds of plants, attend to *individuals* as Craig Foster did. For plants are far from a uniform green mass.

The habits of vines

If we take this idea of the diffused consciousness of an octopus being not unlike the extended awareness of a climbing plant, it gives us an imaginative route into how such organisms might experience the world. I have spent a not inconsiderable amount of time in the worlds of vines, if not always in their natural habitats. Their far-out lifestyles make them excellent subjects when seeking to understand plant experience. Everything they do is so evident in the ways they grow and move, as if their forms trace the history of their experiences. One reason for this is that they have a pressing aim: to find a support to climb up, which in a wild situation is usually a larger plant such as a tree. And they use a wide variety of strategies to detect potential hosts. Some do it mechanically, circumnutating to find a target and then coiling around supports after making physical contact. Some detect certain airborne compounds that hosts emit and make a beeline for them. Others detect different colours of light or head towards looming shadows which might indicate where a support can be found.

In *The Movement and Habits of Climbing Plants*, Darwin gave his thoughts on why vines were so varied in their natures:

Plants become climbers, in order, it may be presumed, to reach the light, and to expose a large surface of leaves to its action and to that of the free air. This is effected by climbers with wonderfully little expenditure of organized matter, in comparison with trees, which have to support a load of heavy branches by a massive trunk. Hence, no doubt, it arises that there are in all quarters of the world so many climbing plants belonging to so many different orders.

Climbing plants come from many and varied lineages, all attempting to cheat the system using different tools and in very different individual situations. For example, dodder is especially driven to locate suitable hosts. Having no chlorophyll, this parasitic vine cannot make its own food. Other plants are not only supports, they are prey. Dodder moves its delicate tendrils around and samples the surroundings with a special ability. The plants that dodder might want to parasitise produce myriad airborne compounds, such as ethylene, which the dodder is able to analyse. These chemicals provide valuable cues as to where the hosts are located.[9]

If you record a dodder with time-lapse photography, it clearly shows how the dodder follows these chemical trails with intent, not unlike worker ants following the trails left by other ants when seeking food. As dodder grows and approaches its target, perhaps a tomato plant it has just sniffed out, the pattern of movement changes. From a snaking, exploratory growth, it changes to a direct approach to the target. Once a host has been clasped, dodder twines itself around its stems, penetrates the vascular system and sucks the nutrients out. From a sommelier sampling and analysing the fine cocktail of chemicals in its surroundings, the dodder becomes a slow-motion vampire.

Right from a seedling, dodder can distinguish the chemicals of different plant species, and between plants that are full of nutrients and those that are wasting away. It can do this without the help of an olfactory system, as well as choosing direction and rate of growth to the preferred target. In fact, seedlings have such small energy stores that they must find a target rapidly or they will die. If a dodder has started to grow towards a host which seems of low quality and senses another, more appetising one nearby, it will change direction and head for the more appealing option. It will always choose a tomato plant over a wheat plant, and will grow much faster towards it. Whereas if wheat is the only option, dodder will grow towards it with apparently little enthusiasm—more slowly and with fewer tendrils. This, it seems, is a trick that wheat can pull on the dodder: it produces a chemical which tomatoes exude when they are nutrient stressed and have little to offer the parasite, masking itself behind this repellent scent. As the dodder explores its surroundings, growing to gather the scents of potential hosts around it, the wheat plays a game of chemical hide and seek, creating a volatile mask to keep itself out of harm's way.[10]

Airborne messages are not the only focus of vines seeking arrangements. The seedlings of one tropical climber, *Monstera gigantea*, have been found to be drawn to dark shapes when they first start growing. This counterintuitive habit has been called "skototropism."[11] But it makes perfect sense for forest climbers: the tree trunks they will wish to ascend are dark. As they grow higher, the climbers switch to seeking out the light and begin decking themselves out with leaves for photosynthesis. Different species of vines seem to associate with specific tree species in the forest, so there is something non-random about their selection. There may well be more complexity to their search than simply shadows indicating the presence of a tall object. Other species can have colour preferences too. In one experiment, ivy leaf morning glory, *Ipomoea hederacea*, was given the choice of different colours of poles.[12] Black seemed of little interest. The plants primarily chose to climb up green and yellow poles, and sometimes red and blue. Most of all, they chose to climb maize plants rather than the coloured poles.

This ability to make fine distinctions between colours, to tease out the opportunities that surround them, is a matter of deadly importance to vines. Though growth might seem like a slow way of moving and circumnutation too slow for the eye to see, the way vines proactively seek out their targets can be almost like animals stalking prey. In the early 1960s, a study using Darwin's glass plate technique revealed the complexity of their pattern of navigation.[13] Figure 6 shows how *Passiflora* tendrils track a support placed in different locations as the shrub grows towards it. In less than eight hours, the *Passiflora* tendril repeatedly changes its shape, tracking a support moved to three different locations. Not only does the shrub clearly recognise the support and work to approach it, it practically *chases* it when it is moved around. This is not surprising. Vines that fail to make an ascent are less likely to survive from seedlings and are very unlikely to reproduce. So we wouldn't expect them to simply cast about and find supports by chance. They must make *choices*. In a complex and ever-changing environment full of tough competition, following the crowd might not be a very successful way to operate. Having an individual way of operating could be the way to gain an edge.[14]

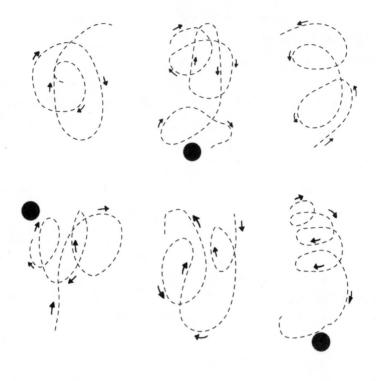

Figure 6

Making meaning

Understanding plant experience is not a simple exercise, though there are very good reasons for thinking we might be able to begin building a picture of it. You can't actually imagine what it is like being just *any* plant, you can only imagine what it is like to be one *specific* plant at a time. Plants show such a variety of sophisticated behaviours that it suggests that the internal states of one plant or another, even in the same situation, are not necessarily always the same. And the behaviour of one individual plant can be incredibly flexible over time too. This variety begs the question: what are the internal states that drive these different behaviours? Can they be ascribed to different subjective states which are highly individual?

Plant science is building a comprehensive picture of the cellular and subcellular nuts-and-bolts of plant behaviour. A solid model of the way plants respond is being constructed.[15] But this tells us little about the subjective experiences of plants, just as creating a time-lapse sequence doesn't really give us an insight into what it's like to *be* a vine growing up a pole; it only makes the action perceptible to human eyes. As plant physiologists detail the underlying processes and mechanisms, we must also explore plants in their entirety and in their environments, as we began to do in the last chapter. Paradoxically, in order to understand what their *interior* worlds are like, we need to focus on their interactions with their surroundings.

This seems to be the one and only way to avoid the risk of "over-interpretation of data, teleology, anthropomorphizing, philosophizing, and wild speculations" that Lincoln Taiz was warning us against. Placing emphasis on the natural setting where behaviour unfolds means that cognition is not something that plants—or indeed animals—can possibly *have*. It is rather something created by the interaction between organism and its environment. Don't think of what's going on inside the organism, but rather how the organism *couples* to its surroundings: for that is where experience is created. Plants must take in their environment and behave in ways determined by how they fit in with it. And this is possibly even more true for the flexible body plans of rooted plants than for roaming animals. Plants tailor their forms and experiences to their environments in a way that animals simply cannot. So if we look closely at *how* they do this, we will be able to begin to understand *why* they do.

There is an interesting realm at the crossroads of biology and semiotics called biosemiotics, which looks at how life and information-creating processes deeply intertwine with one another.[16] It posits that biology is fundamentally all about "meaning-making." Behaviour is geared towards gathering significance from the world, throughout the whole tree of life, even down to the simplest organism. Bacteria such as *E. coli* exchange a molecular language with the environment to help them decide where to move to or avoid. They can swim towards chemical trails that might signal good things and hurry away from ones that might be toxic, and make choices between different options if need be. *Stentor* is a unicellular organism in the kingdom *Protista* that was studied at the turn of the twentieth century,[17] and shown also not to be a basic automaton with hard-wired reflexes. Time-lapse photography shows that it can deploy a gamut of different responses to something it doesn't like—from simple bending and resting to using its cilia and other sophisticated tricks. It appears that *Stentor* gathers information about its environment, tries something in response, monitors the effect, then tries something else if that doesn't work. It makes *choices*, just like our vines, it does not simply give a knee-jerk reaction when provoked.[18]

Biosemiotics leads naturally to the idea that each organism exists in its own particular world. This is made up of the unique dialogue it has with its surroundings: what it knows about them and what it chooses to do as a result. Each type of organism has a different kind of dialogue with its environment, determined by what it needs, how it perceives, and the potential behaviours it has at its disposal. This idea has been called the *Umwelt*, the world at which the individual is the centre.[19] If even unicellular organisms can create their own subjective worlds full of meaning, then surely plants, with all of their incredible complexity and sophistication, must also do so.

There will of course be many differences in how this comes about. Plants are multicellular not unicellular, and they make their own food rather than roaming around to find it—but the essential principle stands. In fact, the term "phytosemiotics" was first introduced in the 1980s to refer to the study of signs in relation to plants.[20] There are many developments needed in semiotic theory to allow it to successfully incorporate plants, and the project is underway. But for now, it is enough to acknowledge that the making of a rich and specific *Umwelt* is absolutely necessary to an organism's ability to survive. They are each the protagonist in their own, sometimes microscopic, drama of survival, deploying their evolved toolkits of sensory abilities and behaviours as they interact with the living and inert world around them. A vine is not simply reacting to the chemicals or dark shapes or physical objects it senses in its surroundings using its particular evolved abilities to detect. It is creating *meaning* from them and deciding on a course of action from the variety of potential behaviours it is capable of. Making an *Umwelt* is essential to plants' ability to successfully handle everything that life throws at them.

Plants-as-animals

There are areas of animal research that we can borrow from to help us in this project. Ethology, the science of animal behaviour, takes this close relationship between organism and environment as its central premise. This might seem like a self-evident idea now, but that is only because of the work of ethologists such as Karl Lorenz, Niko Tinbergen and Jane Goodall. They showed that many animals have much more complex cognitive and social systems than were previously attributed to them, and that they were capable of pleasure and pain. Goodall lived for fifty years with her wild chimpanzees in Gombe Stream National Park in Tanzania. She understood more than most that you cannot make sense of the behaviour of an animal when it is detached from its surroundings. As a result, her findings are truly inspiring. Yet, she and the two other superstars of ethology—Dian Fossey studying gorillas and Biruté Galdikas studying orangutans—were all accused of grave anthropomorphism.[21] Only later was it accepted that empathy plays an important role in understanding the interior worlds of other species, and that to deny it is to wilfully miss something crucial. Just as Paul Churchland implied, we must *imagine* ourselves into different frameworks, just as we imagine ourselves into others' shoes in our dealings with other humans.

This kind of ecological psychology has many useful ideas for the study of plant behaviour. We can think of plants as animals in order to apply ethological ideas, as long as we don't take it too far. One of the major principles of ecological psychology is the idea that animals perceive *affordances*, a word coined by our friend Gibson. Though it really needs replacing, nobody has yet come up with a better one. Here's how he defined it:

> The affordances of the environment are what it offers the animal, what it provides or furnishes, either for good or ill. The verb to afford is found in the dictionary, but the noun affordance is not. I have made it up. I mean by it something that refers to both the environment and the animal in a way that no existing term does. It implies the complementarity of the animal and the environment.[22]

The environment affords possibilities of interaction or "opportunities for behaviour." We may say that the animal's direct environment *offers* resources to act upon, and organisms are on the lookout for these opportunities. Animals find things they can reach, kick, climb, grasp, and so on. Different agents perceive different affordances because their bodies, behaviours and needs are different. The stairs at home afford my young son and me different possibilities of interaction (climb-ability) because my legs are longer. We interact with them differently. I can stride up them quickly, while he has to adopt more of a mountain-climber approach on all fours, summiting each step one by one in a more effortful way. My arms might offer an attractive fast-track route to the top.

To get a concrete idea of what affordances actually are, and how they differ between species, see the figure below, which illustrates some of the widely different ways that different organisms may perceive the same object—for example, a stone. An adult human may perceive the affordance of throwing it, while a mouse might perceive the affordance to hide behind it, and a cat may perceive the affordance of hiding prey.

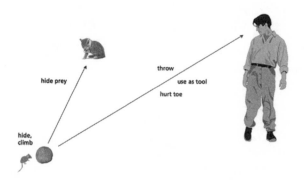

Figure 7

Plants also perceive their surroundings in terms of how they can use them. A vine perceives a support as affording climbing, while the human who put it there might see it as a component for a useful structure, or a butterfly might see it as a well-appointed place to perch. Organisms perceive objects in terms of how they relate to them. They see the possibilities that these objects offer. A human, a cat and a mouse don't see a stone, they see a "thing that can be utilised" in their respective ways, and a vine doesn't perceive a pole, it perceives the possibility of climbing it. These organisms don't detect objects *per se*, they sense the various *possibilities* that objects offer.

An affordance only makes sense to a particular agent. Vines evolved to detect the opportunities to climb that the local environment affords. If a potential support turns out to be far too wide to make it usefully climbable, they might even reject it, twining around themselves instead of attempting an impossible and costly ascent. Non-climbing plants will not perceive such affordances, in the same way that the mouse doesn't perceive the affordance of throwing a stone the size of a human fist. In fact, climbing beans perceive the affordance of poles, but not any pole affords twining—only those of an appropriate size (see Figure 8). The size of the vine, the type of tendril, the properties of the surface of the support—it's all part of the equation. What truly matters is not what's going on *inside* the organism so much as the coupling *between* the organism and its environment.

Figure 8

Tuning in

Though we have used the computer metaphor to understand processing in brains, there might be a different kind of technology
that offers a more subtle route to understanding how affordances
might create experience. Gibson came up with the "resonance
model" to describe the coupling of organism and environment
like that between a radio transmitter and receiver.[23] When radio
waves are sent out from a station, a specific kind of carrier wave
is used. It has a combination of particular frequencies, ampli-
tudes and other characteristics that are unlike those used by any
other station. This signal is then "modulated," which means it
is converted into transmissible information which is set out into
space by transmitting antennae. The signal carries information
about the source, just as different pebbles dropped in water
would create different kinds of ripples on the surface depending on their size, their shape, and how they were dropped. The
sound created by a footstep, light reflected off a trunk's surface,
or chemicals emitted by a tomato plant, all radiate out from their
source and dissipate in strength the further away they get.

On the other side, receiving antennae will pick up these signals, but only if they are "tuned in" to the correct characteristics
of the carrier waves being sent out. The signals must resonate
with the receivers.[24] The kinds of surfaces or objects off which
these signals reflect are what modulates the signals, placing them
within or outside the range of relevant signals that a receiver can
pick up. These ranges are determined by evolutionary history.[25]
Sensory organs are like receiving antennae, attuned to particular kinds of signals. The eyes of different species are sensitive to
different parts of the light spectrum. We can see a narrow band
between 380 and 700 nanometers; some species can see into the
ultraviolet or infrared ranges.

Take information carried by light, for example. We have camera-type eyes that use retinas to create an image which is then processed by the brain. Some animals have eyes with chambers. Others, including insects, have a vision that works on a very different basis, using compound eyes. Plants, too, can "see" by detecting light and comparing the relative levels of light coming from different directions, but have no need to use image-forming organs.[26] There are many ways to pick up a signal. The exploratory tendrils of vines extend out into space, acting like receiving antennae to collect information about the world around them.

Some signals are so important that they have become hard-wired by evolution. Both humans and plants are sensitive to the direction of gravity.[27] We recognise the direction of a sound or sudden movements in our peripheral vision. Plants can sense the moistness of soil or the direction light is coming from. But tuning into other frequencies must be actively sought out or even learnt. We might have to put some effort in to tell the temperature of an object or the texture of its surface by touching it. We learn to finely judge balance when riding a bike or intuit the speed at which we are driving in a car. Plants might have to press through the soil with their roots to get a reading on the nutrient gradients in different directions or circumnutate with their shoots to seek for the solid touch of a support. When we think about cognition in this way, it becomes less about a computer-like bank of memory, and more about a dynamic set of continual interactions with the environment, which shape the signal receivers just as they shape the signals that they are sensitive to and are available to them. The growth of plants into their environment becomes a goal-directed quest for meaning and the opportunities it throws up, which are weighed and pursued accordingly.

Phytopersonalities

The experiences of each plant are shaped by the closely woven interaction between its particular physicality and the opportunities in its surroundings. Each individual creates its own personal *Umwelt*. The experience of one plant is not the same as that of another. And this goes both ways: one plant will not behave in the same way as another might in the same circumstances. We have only just begun to detect these differences. And if we draw all of these ideas together, it looks like plants might have something that we could call *personalities*. This word sits a little awkwardly in relation to non-*person* organisms. But it is the best approximation that we have while we build our understanding of what lies underneath the differences between individuals. Whether it is quite the right term or not hardly detracts from the fact that it represents an invaluable central concept from which to explore further.

The idea that even other animals might have personalities, or display consistent behavioural differences over time, is still quite new. The added difficulty is that we can hardly give them a Myers-Briggs test to work out their personality traits. But some researchers are intent on demonstrating that they do have personalities, despite the lack of formal frameworks for doing so. One group from the University of California gave golden-mantled ground squirrels a series of personality tests with mirrors and traps. They found that some squirrels were more bolshy and active, aggressively seeking out perches and generally getting the most out of their environment. Others were more timid and reserved—and these tended to come off worse when there was a problem to solve, or a conflict.[28] Likewise, in 2021, researchers from the University of Wyoming lured Asian and African elephants into personality tests using rewards of marshmallows.

They gave them a classic "trap-tube" task, often given to primates, along with a series of other tests, and explored whether the speed at which they learned to solve these tasks was related to personality characteristics such as defiance, sociability or aggression. They found that aggression and being active helped in solving tasks but didn't affect learning itself.[29] Studies like these point to the ecological importance of personality: different behavioural tendencies over time might have quite significant effects on how well an individual does.

Not surprisingly, very little attention has been given to the possibility of plant personalities. But we are starting to investigate. Some, such as the mimosa, are easier to read than others. It seems that different mimosa plants have quite individual responses to danger signals, for example. One study measured the time for which plants folded their leaves in response to danger, across a large number of individuals. This folding time was given the rather endearing label "hiding time." And different mimosas had very different individual preferences for how long they hid. Experiments were carried out with putting the plants in different conditions, to see whether, like animals, the plants made a different risk assessment when they were stressed versus when things were going well. Plants kept out of the sunlight for many hours, for example, would only "hide" for a short time at the hint of danger. They had to take the chance of being eaten when they hadn't been able to photosynthesise enough food for a long time. Whereas plants that had plentiful sunlight exposure would stay folded for much longer—they could afford to play things very safe, being replete with energy stores. The researchers concluded that the state of the plants accounted for much of the variety of their hiding times. But the rest of the variation was down to individual preference.[30]

Some of the most dramatic differences in personality are the ones that have been shaped by human domestication of other species. If you compare a mountain goat practically scaling a sheer cliff wall to escape a predator to a sheep munching blankly in a field, you immediately see the difference. This can occur in an incredibly short space of time: we domesticated dogs from wolves over 33,000 years, transforming their anatomy and behaviour to become companions for humans. But a striking transformation was enacted in silver foxes in about forty generations in a 1950s experiment by the Soviet zoologist Dmitry Belyaev. Selectively breeding the tamest individuals produced something akin to the floppy-eared appearance and sociability seen in domestic dog breeds.[31] Likewise, we have domesticated plants as food, raw materials and ornaments ever since we formed settled communities 10,000 years ago. Take florists' gloxinia, for example. We've been domesticating it for over 200 years, and in that time it's undergone a paradoxical change: its genetic variability has decreased, while the variation in appearance has become astonishing. Despite gloxinia's relatively small genome, it now has the myriad colours and shapes of a species like the snapdragon, which we've cultivated for 2,000 years.[32]

The wild vines I went searching for in Mauritius, back in the Introduction, have not been altered by humans for human interests. Domestication has dramatic effects on vines, not only in making them produce better fruits or flowers for human delectation. They start to grow shorter distances between the branching points on their stems; they become more sluggish and less canny. These plants are essentially dwarves, and perceive different affordances from wild relatives as a result. In comparison with their wild forbears, they are no longer able to cast so widely and find a support with such efficiency. They sometimes lose the complex microbiome that exists in the roots of wild plants and gives them access to essential nutrients.

But this is hardly a problem, because they are given poles and trellises to grow up, conveniently set right next to where they emerge from the ground. They are given fertiliser and other soil enhancements to make up for the lack of root symbioses. These changes can make them much easier to cultivate and harvest from, but they would not do well in the wild. We use domesticated climbing beans for our experiments in MINT Lab, so we also need to observe wild beans that are full of impetus and able to circumnutate in a wide circle. Considering how fascinating the time-lapse findings are with domestic vines, imagine what we would see if we time-lapsed wild vines. Some cultivated plants do manage to return to the wild. They go feral, re-acquiring some of the growth patterns and characteristics of their wild ancestors, though not by reversing the genetic changes that happened through the domestication process. They are new, unique breeds, sometimes escaping with artificial genetic modifications that make them resistant to herbicides. These doughty strains evade the tyranny of human-guided artificial selection to take their chances under natural selection. They have individual experiences of an entirely new kind, as the products of human selection newly encountering the challenges of the world.[33]

Anthropophilia

In a 1991 lecture at the Royal Institution in London, Richard Dawkins described an image of the "ultraviolet garden:" the garden perceived from the perspective of the interactions between cultivated ornamental plants and their pollinators. We might think that the beautiful blooms we grow are for our enjoyment only, the objects of human anthophilia or love of flowers. But they have a far longer history to which we are only a recent addition. We've already discussed the interactions between flowers and pollinators. Much of their discourse, however, happens in the ultraviolet spectrum, which we cannot detect. Flowers lay down ultraviolet markings to guide their UV-sensitive pollinators into their interiors. Each side sees the relationship differently: bees see the affordance of landing, while flowers perceive "guided missiles for firing pollen from one flower to another." And they have shaped each other over evolutionary time through the exploitation of these affordances. As Dawkins summed up:

> Flowers use bees, and bees use flowers. Both sides in the partnership have been shaped by the other. Both sides, in a way, have been domesticated, cultivated, by the other. The ultraviolet garden is a two-way garden. The bees cultivate the flowers for their purposes. And the flowers domesticate the bees for theirs.

We mustn't be so conceited as to see ourselves as above such reciprocity. Even thinking of ourselves as doing all the domesticating of plants is returning to our usual anthropocentric ways. Such habits of mind are very difficult to avoid. If you consider what has happened to the plants that have received our agricultural attentions, they have been spread around the globe, grown in specially manicured areas of land, protected from pests, fed valuable nitrates, and their plant competitors beaten off with herbicides. Modern wheat or corn varieties have done very well in comparison to the thrifty wild counterparts from which they were bred. They might have become complacent and dim, but that is because they can afford to be. Their human keepers look after their interests. But perhaps we are actually being cultivated by them.

We might consider that we were not the only actors here. Maybe the plants disposed themselves to domestication, for in that direction lay an easy life and unimaginable proliferation. One trait, the production of juicy fruits that we love to eat, in fact began as a bargaining tool with mobile animals to help plants reproduce: "I'll give you some nourishment, if you carry my seeds away and deposit them somewhere with a rich pile of dung to fertilise their early growth." It became the hook which drew us to care for them, cultivate them, and breed them to produce even bigger and more appetising fruits. A recent study shows just how consistent the patterns of plant domestication over time and across geographical regions have been. Perhaps their malleable forms and amenable dispositions predisposed them to infiltrating our lives and becoming mainstays of human existence. Plants might have been shaped by their own anthropophilic tendencies, working the affordances of a domestic arrangement to their considerable advantage.[34]

Few plants have done so well in playing this game as our houseplants. Not only do they have humans cultivating them, they are manicured, fed and watered—sequestered away from competitors, predators and parasites. For many living in small urban households, plants serve as replacements for the more demanding animal pet variety. Were we to consider it, we might imagine that the individual experiences of our pet plants are rather extraordinary compared to the majority of photosynthetic lifeforms. No other plants have such attention lavished on them and simultaneously so much agency taken away from them, not to mention the rooted isolation in their individual containers. If we can look beyond the basic care of trying to just keep the plants in our homes alive, we might be able to consider what they are experiencing as members of our households. We can imagine ourselves in the diffused consciousness of their growing forms, the unusual chemical signals they pick up, the strange light patterns, the artificial landscapes, and of course what it is like for them living perpetually with our noisy, chaotic activity. We might feel their presence as company rather than mere ornamentation.

CHAPTER EIGHT

PLANT LIBERATION

Delving into the inner worlds of plants—or octopuses or bacteria—is not only a subtle exercise. It is also one which could have profound implications for the way we see the world and how we choose to exist in it. We've explored the complex ways plants collect and use information from their environments, the smart behaviours they are capable of, and the complex relationships they have with other organisms around them. We've considered what it is like to *be* a plant. And the answer to that seems to be far more profound than is comfortable. So, where does all of this leave us? Apparently with quite a pressing ethical dilemma. When I give talks on my work to general audiences, the first hands to shoot up to ask questions invariably belong to vegans and vegetarians who have had their ethical frameworks thoroughly shaken by what I've said. If plants are meant to be ethically "safe" to consume because they can't suffer like animals can, then the burgeoning possibility that plants have subjective experiences very much topples zoocentric claims to moral high ground.

This possibility needs examining closely, before we jump to conclusions. But we most probably have a great deal of thinking to do about our stance on our treatment of not only other animals but many other kinds of lifeforms. Though not everyone agrees, of course. In 2020, the stalwart critics of our work at MINT Lab published some essential objections to the possibility that plants have subjective experiences. Their argument is twofold. The first thread strikes at the link between integrated responses and sentience, arguing that "the capacity to process environmental information for adaptive behaviour and subjective awareness of the environment are two different things;" the latter, they propose, requires neuronal systems with epicentres that are something like brains. Their second strand of argument asserts that the evolution of consciousness is unnecessary for plants, that hardwired adaptations are sufficient to service photosynthetic lifestyles. They suggest that "instead of subjective consciousness, plants evolved adaptive behaviour that is genetically determined by natural selection and epigenetically determined by environmental factors."[1]

We welcome these objections, because they provide a critical gauntlet which our ideas must run in order to prove that they are well-founded. And we can answer both confidently. First, even if "consciousness," as understood in vertebrates, is generated by complex neuronal systems, there is no objective way of knowing that subjective experience has not evolved with entirely different kinds of hardware in other organisms. We have no evidence to conclude that no brain means no awareness. Second, the work we have done to understand plant behaviour makes it very difficult to reduce it to mere adaptation underpinned by genes and environmental influences. The behaviour we see is far too goal-directed and flexible for that. Even if we take a very fundamental definition of consciousness—the presence of "feelings, subjective states, a primitive awareness of events, including awareness of internal states," we cannot yet know if plants are conscious. But we also cannot assume that they are not.[2]

We may not be able to perfectly execute the "recalibration"

exercise of perception that Paul Churchland encouraged. It is impossible to actually know what it is like to *be* another person—even a long-term spouse—or a bat, or any other animal, never mind radically different organisms. That doesn't really matter when we are thinking about implications. If there is even a chance that plants are sentient, then we still need to consider the ethical consequences of such a possibility. For this consideration, I see a dodder or a pea plant as somewhat akin to a "locked-in syndrome patient," those who exist outwardly in a vegetal state but who have an awareness of what is going on, and an inaccessible internal experience. They cannot communicate with those around them to express their feelings and needs, except via blinks and vertical eye movements. So any consideration that is paid to their wellbeing is based purely on speculation and ethical choice. What rights should they have?

This is just where we are with plants: they may have the ability to suffer as part of an internal experience. We don't know, they cannot tell us, and we haven't yet deployed the scientific tools to find out. But we need to consider what it means if they do. To do that, we might need to re-examine our assumptions about where consciousness comes from, to concepts that might span across widely different groups of organisms. Then, we need to decide what kinds of consciousness we are going to value.

Emotional behaviour

When we consider what our consciousness is, what it is like to *be* a human, what might spring to mind is the ability to form abstract thoughts or conceive of ideas. But is this really just a veneer on top of the internal drivers of what goes on inside us? We cannot escape the fact that a large proportion of our behaviour is emotionally driven: whether that be behaviours that

express emotional states—such as laughing, crying or frown-
ing—or behaviours that are not about communication but have
emotional underpinnings. These "feelings" are mental states
representing groups of physiological functions that usually have
defined behavioural purposes. So we could say that emotional
behaviours are those that express internal states, which themselves are adaptive. Fear, anger, affection and other emotions are
essential drives motivating our interactions with the world, and
possibly those of other organisms too. This emotional impetus is
also what connects us most closely to other species.

In 1872 Charles Darwin published *The Expression of the
Emotions in Man and Animals*, which might today have been
called *The Expression of the Emotions in* Humans *and* Other
Animals. He described how even "insects express anger, terror,
jealousy and love, by their stridulation." Stridulation is how
insects such as crickets and grasshoppers produce rasping songs,
by rubbing brittle exoskeletal surfaces together. Whether or not
these sounds are really emoting jealousy and love, the essential
recognition of emotive force in animal behaviours was a dramatic
contrast to the existing idea of animals as automatons. Darwin
saw emotions even in "lower" animals such as insects that were
so akin to those that we ourselves experience that they might be
described with the same language.

Darwin was engaging with a debate that has continued since
the nineteenth century about the relationship between emotions
and the behaviours by which they are expressed. There have
long been questions about whether emotions are specific to the
anatomy of the human brain and human behaviour.[3] Darwin
was one of the first scientists to consider the evolutionary importance of "feelings," beyond being an abstract distinguishing
feature of humanity. Emotion and emotional behaviours, he
argued, evolved for very good reason. They give the capacity
to make rapid, prioritised decisions in response to the demands

of a dangerous environment. We might think of emotions as irrational drives, but there is great value in the phrase "trust your gut." Sometimes emotions, the subjective internal experiences, can help drive complex behaviour in ways that reasoned logic cannot.

Feeling pain

Darwin's perspective not only allowed the emotions of other species to be considered, it made them an important avenue for investigation. The work of ethologists such as Jane Goodall in the twentieth century demonstrated unequivocally that non-human animals can feel pleasure and pain, and have emotional interactions within their complex social structures. Their work raised difficult questions about how much importance we should place on the suffering of other mammals, especially those not so different from us such as the great apes. Many of these debates centre around the infliction of pain, which can be defined as an adverse sensation which stimulates an aversive response.[4] Under any kind of reasonable ethical framework, the infliction of pain should be minimised. But the question of whether or not other animals feel pain has been hotly debated.

In 1975, the Australian philosopher Peter Singer wrote the now-classic work on the ethics of animal treatment—*Animal Liberation*.[5] He relied heavily on the work of ethologists such as Goodall and argued that there are three strong grounds for believing that non-human animals can feel pain. First, the kinds of behaviours that animals exhibit in situations when they might be likely to feel pain. Second, the complex nature of their nervous systems, which would allow the detection of and response to pain. And third, the fact that pain, as an indicator of damage or danger, is extremely useful from an evolutionary perspective.

Singer explicitly excluded plants from his argument, saying that "none of these gives us any reason to believe that plants feel pain." I would take issue with this exclusion: plants show actively avoidant behaviour, they have "nervous"-like systems that can coordinate responses through their bodies, and pain should be no less useful in the evolutionary history of rooted organisms than for those who can run away from it.

But let's start by moving down the evolutionary tree, and seeing how far our assumptions about the experience of pain go. Do fish react to noxious stimuli painlessly, for example? Is that why many people feel more comfortable with fishing than with hunting? How do we know fish don't feel pain? Fish brains have an area called the pallium which share an evolutionary history with the amygdala and hippocampus of mammals, which register fear and pain. The philosopher Brian Key argues that fish cannot feel pain, for to feel pain it is necessary to have a mammalian neocortex. So any experience produced by a different nervous system structure must be something different. Thus, Key suggests, fish merely act in a manner that we *interpret* as distressed. A hooked fish flopping about on the bottom of a boat is really only reacting automatically to oxygen deprivation. Key implies that the thrashing fish has little more awareness than those plastic mechanical singing trophy fish sold in tacky gift shops.[6] The implications of this assumption are exceedingly convenient for the human purpose of guilt-free fish consumption.

Key's argument echoes Descartes' assertion, which we touched on in Chapter Five: that non-human beings lack souls and intellects, so they are nothing more than machines. On the strength of this assumption, Descartes and his followers would carry out horrific vivisection experiments on dogs, nailing them up by their paws. "How could anyone be so heartless?" you might wonder. The answer is: they were inured to the animals' expressions of pain because they believed them nothing more than automatons.

Outrage aside, it is deeply ironic that the prioritisation of intellect made the Cartesians act like unfeeling monsters. Yet many of us who might avoid eating mammals and birds on account of their suffering under intensive farming would still happily eat fish that had been killed in the suffocating embrace of a trawler net.

The denial that non-human organisms have feeling lasted well beyond Descartes and his followers. Even Richard Dawkins, for example, who has influenced public perception of evolutionary theory perhaps more than any other person in modern times, argues that "a bat is a machine, whose internal electronics are so wired up that its wing muscles cause it to home in on insects, as an unconscious guided missile homes in on an aeroplane."[7] But we should be prepared to be seen as barbaric by future generations if we deny the existence and importance of other species' pain. Evidence is emerging that fish do indeed have sentience.[8] Goldfish have been shown to be capable of learning; other fish certainly have a sensory awareness of elements of their environment, such as object colour, just like our pole-climbing plants. And it would seem to make little sense for fish to have evolved the ability to react to colours such as red without forming an internal representation of "red," which equates to a form of sentience.[9] The question is, can findings in goldfish be generalised to hake, sprat or tuna—even beyond, to entirely different groups? If complex networks of neurones mediate behaviour and emotions in humans and particular non-human animals, it is an open question whether the same holds across phyla.

However, even if we cannot specify the similarity between a particular experience of "pain" among mammals and other organisms, we can broaden our concern to a more general idea of "suffering," which might be more easily applicable to plants. There is an intriguing link between plant behaviour and emotions. After all, in the case of animals, feelings get coordinated in the brain stem. Many of the chemicals that control behaviour and emotions in humans and other animals are also synthesised or have analogues in plants: auxin for example is chemically very similar to neurotransmitters such as serotonin, dopamine and adrenaline. Melatonin, the chemical which regulates our own circadian rhythms—the internal clock entrained by the cycling of day and night outside—also seems to do the same for plants.[10] These substances are expensive to produce, so it would make no evolutionary sense to manufacture them without purpose. And the more we find out about the function of these molecules in plants, the more similar their use appears to be to that in animals.

Some of these chemicals are only produced in situations when plants are stressed or injured. Plants make many substances that have painkilling or anaesthetic effects, such as ethylene. Ethylene seems to be an important stress signal in bacteria, fungi and lichens as well: its message bridges large swathes of the evolutionary tree. We don't know that these molecules act as painkillers *per se* in plants, but given that they are created in stressful situations, there is reason to believe that they serve to relieve suffering. We can even measure plant stress directly now using nanosensors. Carbon nanotubes embedded in plant leaves detect ethylene and other signals that are produced when leaves are damaged or drought stressed, and can visualise plant distress in real time. This information can even be streamed directly to our phones.[11]

In the case of plants, our working hypothesis should be that they exploit coordinated physiological activities in order to deal with demanding environments. As in the case of animals, inner states help them to create priorities, organizing the demands of life in order of how urgently they should be responded to. From an evolutionary standpoint, the ability to perceive pain or to suffer in some way is essential. In a dangerous and constantly shifting world, organisms must be able to respond to negative events. These events must be represented with some kind of internal state or feeling in order to *motivate* a response, which figures as some kind of basic sense of awareness.

Cellular consciousness

The difficulty with shifting our perspectives on consciousness to include organisms that we perceive as lower, simpler, or even practically inanimate, is that we take a "top-down" view of consciousness. The intellectual froth churned up from the human brain is what we see as distinguishing us from other species. Instead, we can extrapolate from the molecular similarities between animals and plants to build an emotion-led picture of consciousness—one which defines consciousness primarily as some kind of *awareness*. Then, we might arrive at a very different kind of approach, to a bottom-up view of consciousness. We could see subjective awareness as an integral feature of life, however simple or small.

The eighteenth-century French philosopher Julien Offray de La Mettrie described in *Man, A Plant* the continuities between humans and plants, particularly that both have minds, even if those of plants were "infinitely smaller." Nearly 150 years later, in *The Sagacity and Morality of Plants*, the naturalist John Taylor described how the work of Darwin, Alfred Russel Wallace and other scientists on plants implicitly suggested that they had an inherent intelligence or purpose. He argued that they knew that "perhaps there can be no life, animal or vegetable, unaccompanied by consciousness! The minutest animalcule, lowest placed in the scale of animal being, displays a consciousness of external surroundings as simple and elementary as its own structure," and posited that the "Vegetable psychology" they practised might soon lead to a future in which it was accepted that "there can be no life absolutely without psychological action—that the latter is the result of the former."[12]

A relatively new theory which goes a great way to solving the dilemma of how to see consciousness has been developed by cognitive psychologist Arthur Reber at the City University of New York.[13] He has flipped the exploration of consciousness on its head, from an anthropocentric focus pushing its boundaries slowly outwards to include more lifeforms, to an essential perspective. Consciousness is found throughout the tree of life, Reber argues, because experience is inherent to living. In his preface to *The First Minds* he proposes

> that unicellular species like amoebae have minds, though they are very tiny and don't do much, that protozoa perceive the world about them and think, though their thoughts are limited in scope and not terribly interesting, that bacteria communicate with each other, though the messages are simple and unitary in nature, that sessile eukaryotes like *Stentor roeseli* not only learn, they have minuscule cellular memories and make tactical decisions.[14]

Viewing consciousness in this way, he argues, allows it to become conceptually continuous with evolutionary biology. It makes consciousness into something manifested essentially in the molecular details of cellular life, from which more complex minds may have evolved, rather than an abstract phenomenon impossibly hung on the mind–body problem.

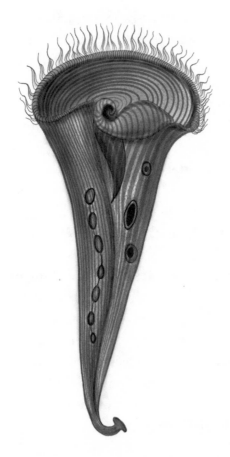

Portrait of the ciliate protozoan Stentor roeseli *that lives worldwide in rivers and bodies of fresh water. After* The Biology of Stentor *by Vance Tartar (New York: Pergamon Press, 1911).*

Under Reber's theory, subjective experience must have existed since cellular life evolved, as a necessity for interacting with the environment. It is not something restricted to a small group of privileged brain-owning organisms. As Reber wrote in a subsequent paper: "a nonsentient organism ... would be an evolutionary dead-end."[15] Even single-celled bacteria do not just "sense" their environments, they *perceive* them, detecting the valences that exist for them *specifically*. The subjective experience of encountering a sugar molecule and the rich vein of fodder it might signify is key to the bacterium's choice to move towards it; the negative effect of being touched by an acidic molecule drives its motion away from danger. Unicellular organisms are motivated by strong urges: to feed, excrete, avoid danger. Their minute lives are full of the passions of the most basic biological needs. Single-celled organisms such as bacteria and amoebae can even learn. Bacteria fed with lactose followed by maltose, each requiring different genes to be activated to be able to deal with them, quickly anticipate the forthcoming change by synthesising enzymes to digest maltose before the maltose has arrived. And should the expected maltose never be brought to the table, they will also switch the anticipation off, just like Pavlov's dogs after the bell rang too many times without the food arriving.[16] If this is what single-celled organisms are capable of, how much further might complex, multicellular ones go?

Oddly though, despite his egalitarian framework, Reber initially excluded plants, making "capacity for locomotion" one of the three essential qualities for possessing the "biological foundations of mind and consciousness." Another, flexible cell walls, likewise pre-emptively shut out plants. The exclusion shows just how deep the vein of zoocentrism runs.[17] Reber recounted some of his initial inspiration for the theory, watching a caterpillar inch across a basil leaf, neatly clipping the rim as it went, actively selecting choice pieces to partake of, rather than mindlessly gnawing whatever came up in front of it. It suggested to him that not only did the caterpillar have a mind, it had some form of consciousness, for only then could it make the most of the complex opportunities in the green landscape it navigated.

I ask in response, what of the basil plant? Predator and prey, parasite and host evolve in tandem, not in isolation. If one has a mind to direct its consumption, why not the other to defend itself? The vibrations created by a caterpillar munching are markedly different from those generated by a passing wind. They must be a different *experience* for the basil plant, shaking its internal world differently and motivating very different responses.[18] Just as a bacterium's experience of an acid molecule might be deeply unpleasant, a plant's encounter with salty soil, or the tearing jaws of a herbivore, might well be too. Darwin himself, when watching earthworms, had similar instincts to Reber which he described in *The Formation of Vegetable Mould*. He saw "how far the worms acted consciously and how much mental power they displayed" in their careful selection of organic matter to plug their burrows. But in his earlier work, *The Power of Movement in Plants*, he showed none of the same zoocentric prejudices.

Reber responded to my objections with aplomb, conceding that research was building an ever-strengthening case for the possibility of plant sentience. He drew from his own premises to support this: "If the Cellular Basis of Consciousness model is correct and prokaryotes are sentient, then by the simple fact that plants evolved later, they should have retained their capacity for sentience."[19]

Evolution is conservative. Once something useful has evolved, and continues to be useful, it is unlikely to be lost, even through the millions of years between plants' single-celled ancestors that absorbed smaller photosynthesising cells and their complex multicellular descendants. But on the same count Reber posed another challenge: sentience is metabolically pricey, and evolution abhors waste. He suggests that the ancestors of plants, relinquishing motility, might at the same time have been able to usefully channel resources away from sentience into a more specialised lifestyle. The question then becomes, do we view "sessile" organisms such as plants as being able to utilise consciousness sufficiently to make that cost worth it? I think that the plentiful arguments we have put forward throughout this book suggest that the fact this is even in question is a problem of our own incapacity to view the dynamism of plants.

Proving experience

How might we go about giving a solid scientific basis to the existence of minimal consciousnesses? Could we explore how plants function from the *inside*, rather than intuiting what is going on from promising external signs such as learning and anticipation? We can imagine ourselves into the minute world of an amoeba, or the cool tendril-tips of a pea plant, but that doesn't count as evidence, it only acts as a heuristic, a guiding premise for how to approach the question.

Reber has suggested some routes for investigating cellular consciousness using genetic methods. He argues that the early life that first "woke up" must have had this sentience inscribed on its genome, sequences which coded for the mechanisms of awareness. He speculates that the new development of the CRISPR gene editing system, taken from a bacterium, which neatly clips sections of genome out or inserts them, could be used to investigate this. If genes that might be involved in consciousness could be identified, and one by one systematically removed and the effects tested, we might determine which genes make the difference between a switched-on *E. coli* and a half-wit bacterium.[20]

Another potential route leads us full circle to the dormant mimosa plants that we met in the Introduction. If we want to prove sentience or awareness, we can perhaps use its opposite to identify it. Organisms like bacteria don't "sleep" in the same way as mammals or other animals with nervous systems do, in clear phases such as short-wave and REM sleep. But they do, like animals and potentially all other organisms, need downtime. Taking a break from the business of living seems to be one of the essential features of being alive.[21] It gives time to repair damage in cells and reset the system. And we have clear evidence of these cycles. Cyanobacteria, photosynthesising unicellular organisms, shift the genes they express at different times in the light–dark cycle, and the cellular mechanisms of many kinds of bacteria also trace day and night.[22] The jellyfish *Cassiopea* shows clear sleep-like states, which might seem unnecessary when you are essentially a floating mass of tendrils, but is very necessary at a cellular level.[23] One study looked at the physiological changes in zebra fish when they were dormant and found that they showed a similar slowing of heart rate and brain activity patterns to those found in the slow-wave sleep and REM sleep of mammals, suggesting that these kinds of reboot states might have evolved in vertebrates over 450 million years ago. There is even evidence that urban fish, like urban humans, have disrupted sleep patterns because of the ever-present illumination of their habitats.[24]

We have already seen how sensitive plants are to anaesthesia. They also have circadian rhythms that can be "jetlagged" artificially in the lab, simply by keeping the plants indoors with lamps that can be switched on and off for periods out of sync with the light outside. Plant systems are so sensitive to light changes that on 21 August 2017, a solar eclipse in Wyoming caused the dominant vegetation, big sagebrush, to go almost dormant, but not quite to the level of night time. Their photosynthetic output for the day was massively reduced, far more so than the loss of solar energy for a brief period could account for. The period of darkness was not predicted by their internal clocks, but the drop in light had a similar soporific effect to that of dimming the lights on human brains.[25] We don't yet quite know what the mechanisms underlying the effects of anaesthesia or drowsiness-inducing darkness on plants are. Some think that they might be important clues to the interior lives of plants. Plant researchers František Baluška and Ken Yokawa argue that since "Anesthesia in humans induces a loss of awareness," there could be an analogous loss of awareness in plants.[26] Anaesthetics might not only take away a mimosa's ability to fold its leaves, or a Venus flytrap's ability to snap shut: they might temporarily shut down these plants' sentience.

As an anaesthetic takes hold we might ask: what exactly is being closed off? We cannot feel all the unique minutiae which make up the experience of being another human. So we cannot really know what closes down when a mimosa stops responding under anaesthesis. Internal experience is infinitely individual, just as the possibilities that an organism perceives in its surroundings are particular to its unique being. Perhaps we can use the fact that organisms can be "put to sleep" to create a different kind of model of conscious experience.

Giulio Tononi at the University of Wisconsin–Madison, together with Christof Koch and colleagues, have developed a theory which does just that.* They have called it Integrated Information Theory (IIT), based on the idea that subjective experience is made up of key essential properties that are woven together inextricably.[27] It is unique to the individual it belongs to, and exists only within them. It is structured by different components—sounds, sights, tactility—but also exists only when all are present at once. Watching a caterpillar crawl around on a basil leaf on a sunny day with a soft breeze brushing your skin, for example, is not the same experience if any of those elements are missing. The shape of the letters you are reading, the colour of the paper, its texture on your fingers, the meaning of the words and the sounds you hear around you as you read

---

* Koch runs the Allen Institute for Brain Science in Seattle.

are all closely bound together as part of a single experience. If something changes, the experience ceases to be, and becomes a different one.

IIT pictures consciousness as the result of systems that are able to integrate these different aspects. They make an experience into more than the sum of its parts. The more complex the system, the better able it is to integrate elements into a unified whole which cannot be reduced to the individual components. So the power to integrate, to combine information into something entirely unique, becomes a measure of consciousness. Though IIT has been developed in the realm of neuroscience, its premises and implications are not inherently neurocentric. The idea holds whether we are talking about neuronal networks or not. It can be silicone chips, neurones, single-cell membranes, phloem tissue: any kind of system that can achieve the transmutation of nuggets of input into a coherent internal experience can create consciousness at some level.[28] Not only that, but creating a measure of consciousness that can be generated in multiple different kinds of systems allows us to make practical moves towards investigating it beyond the mammalian brain. If we can measure integration to get at subjective experience in an objective way, we don't need to rely on measures of *brain* activity.

Zap and zip

We have started to develop a branch of IIT research geared towards the plant world, which we have named PLANT-IIT, and have made some predictions for what we might uncover. In animals, the electrical activity of nerves integrated in brains and ganglia is likely to produce consciousness. In plants, vascular cells form extended, interconnected bundles with nerve-like functions.[29] There is also probably a wide diversity of levels of sentience among plant species and individuals: the premise of IIT, after all, is that experience is ultimately unique. PLANT-IIT might be a way to begin sifting out what these differences are.

This starting point offers some ways of beginning to investigate plant consciousness. If vascular systems coordinate plant consciousness, then tracking the changing states of the vascular tissue might offer insights into its workings. Magnetic Resonance Imaging (MRI) or Positron Emission Tomography (PET) methods that are used to give a real-time picture of the nervous systems of humans and other animals could be adapted to map the fluctuations in plant vascular systems.[30] Plant-PET and plant-MRI, along with other non-invasive techniques, would give us an insight into just how these vascular systems operate and interact in the plant, perhaps revealing hierarchies of organisation like those in animal nervous systems that are invisible to us now. We could test how the effects of anaesthesia manifest in the way that vascular systems operate. If it seems to cause a breakdown of the integration ability of these networks, we could visualise the temporary shutdown of experience. This work would give us a clue to the structure of possible plant consciousness. All we need is to build the specialised equipment that could allow us to do this.

Giulio Tononi and his colleague Marcello Massimini, now at the University of Milan in Italy, developed a method together in the early 2000s which has become a gold standard for detecting consciousness.[31] It is known by the rather light-hearted name of "zap and zip," but what it uncovers is both revelatory and at times deeply saddening. The "zap" is a magnetic pulse sent through a patient's skull by a coil of wire pressed against the scalp, inducing an electrical pulse in the adjacent neurones which ripples through the connected neurones around them. This is called "perturbation." The "zip" is the collection and compression of the data of the pattern of perturbation from a network of electroencephalogram (EEG) sensors arranged around the head. The more complex the pattern, the larger the "zip" file that results, and the more conscious a patient is. Unconscious or anaesthetised patients show simple, regular perturbation patterns and have low zip values below 0.31. Conscious patients show rippling and shifting patterns with zip values between 0.31 and 0.70. They tested this method on patients in vegetative states, and, to their dismay, nearly a quarter of them had values that suggested the patients were conscious but could not indicate it in any way. They were active minds trapped in inert bodies.

Plants, not unlike these locked-in patients, might well have significant conscious experience, although there is no way for us to intuit it nor for them to communicate it to us. But "zap and zip" might be the first, tentative bridge between our worlds. Rather than applying magnetic "zaps" to crania, we could apply them to regions on the plant phloem, and watch the images of excitation snaking out through the vascular system. This would show us what patterns of electrical activation happen in a plant, how the scaffolding of its internal communication works. The prediction would be that "aware" plants would have, like conscious human patients, more complex and extensive patterns of resonance passing through their vascular systems. Those in a vegetative state, as it were, would show only simple and localised wave patterns. IIT theory predicts that the more awake a plant is, the more complex its consciousness, the more power it has to integrate information between the data-collecting hubs throughout its body. We might be able to begin to liberate plants from their seemingly locked-in state.

Phytoethics

There are numerous enticing threads that might lead to the possibility of phytosentience—whether they be experimental evidence that plants are more than what we have assumed, new frameworks for thinking about consciousness across life, or techniques for probing consciousness directly. But they haven't yielded any firm answers yet. The question of consciousness itself has wracked philosophical and scientific minds for millennia, and we are only just beginning to explore the possibility that there is a verdant addition to this mystery. But what we have seen so far should give us pause for thought. Sentience makes sense for life, as an essential underpinning to the business of living. And it is very unlikely that plants are not far more aware than we intuitively assume.

Arguments regarding our treatment of other organisms have often centred on the infliction of pain. In *Animal Liberation*, Peter Singer argued that we should work to minimise suffering in other animals. It was a watershed in our attitudes towards other species, but he was not the first to make these arguments—it took a long time for ethical considerations held by a few to become more widely accepted. Even hundreds of years ago, Leonardo da Vinci practised vegetarianism in order to avoid the infliction of pain on animals. He thought that they had been given the ability to feel pain in order to detect if they damaged themselves as they moved around. Just as Singer ignored "lower" animals like insects or molluscs, though, Leonardo thought that pain was unnecessary in plants. He argued that, being immobile and unlikely to bump into things, plants had no use for pain, so he need not worry about them. The pain of plants, needless to say, figured in neither of their ethical frameworks.

Even the strong chance that plants are sentient means we need to do some serious thinking. We can no longer turn a blind eye to the ethical implications of our interactions with them. Rather than pain, the wider phenomenon of greater importance is: "how aware is this creature?" If an organism has awareness, then our treatment of it has implications for its suffering. And, if we are to think of ourselves as ethical creatures, we must consider the suffering of other organisms. From the evidence we have, from the behavioural to the physiological minutiae, this almost certainly includes plants. We needn't know exactly what it is "like to be a plant" in order to care about inflicting suffering on them. But what does this mean: how can we minimise the suffering of beings we barely understand and rely on in so many fundamental ways? Even the considerable mind of da Vinci might have had trouble unravelling that question.

Reber suggests solving this problem from a utilitarian position. The *Stanford Encyclopedia of Philosophy* defines utilitarianism as: "the view that the morally right action is the action that produces the most good. In the utilitarian view one ought to maximize the overall good—that is, consider the good of others as well as one's own good." Classic utilitarianism aimed to create "the maximum amount of good for the greatest number," irrespective of individual identity. My good is of no more value than anyone else's good. Traditionally, this referred to humans, but we might need to make a drastic change. The "good for the greatest number" might have to become "number of species" across the tree of life, or number of organisms of any kind. Not just humans. Though how far down the chain of simplicity we might go is not yet clear. Do we worry about cyanobacteria and amoebae? Or only multicellular beings? Organisms with defined brains, or anything that might have consciousnesses created by other systems? Some argue that we must go even further in the case of animals. Dr Tom Regan, the "father of animal rights," argued in his 1983 book *The Case for Animal Rights* that non-human animals had a right not to be seen or treated as commodities or resources, and that the institutions that treated them as such should be abolished. Few are anywhere close to arguing this on behalf of plants.

Put together with new, objective measures such as IIT for detecting consciousness, this fresh way of applying such an ethical principle can begin to have legs—or roots. If we think of the wild vines from the Introduction that I went to catalogue in Mauritius and compare them with vines in the lab, their experiences could not be more different. The wild vines inhabited a humid world of intertwined branches and trunks offering opportunities for climbing ever higher towards the strong equatorial sun, rooted in rich humus teeming with micro-organisms. Their "captive" lab-bound relatives lead strip-lit lives in root-constraining pots of sterile compost, circling across bare soil to seek a single pole to grasp. When we think of plants like this, we might want to start asking the same kinds of questions that we ask about our treatment of animals in agriculture and scientific research. How can we reduce the number of individuals we use? What responsibilities do we have for their wellbeing? Is it justified to bring wild organisms into captivity? Arguably, we do not know how to manage them sustainably in the wild, never mind keep them happily in captivity, and the idea that we do is pure hubris, part of a long history of controlling and dominating other species.[32]

These questions are all open for discussion, just like the questions of whether plants are conscious, and what the nature of this consciousness is like. But, in a similar way to toddlers emerging from their self-centred existences, waking up to the fact that other people have interior lives, motives, desires and needs, we are beginning to see that the human mind does not encompass the world. It is only one, very particular way of experiencing it, among infinite numbers of others. Our minds, without doubt, have the ability, if not yet the knowledge, to consider others unlike ours. And we could grant them rights, or at the very least some consideration.

CHAPTER NINE

GREEN ROBOTS

Planting space

We love exploring new worlds. The lure of the uncharted, unseen or unexploited pulls us to invest vast amounts of time, energy and resources into accessing it. As long as we believe it will enrich us, somehow. Yet we can miss the alternate worlds that sit right under our noses: the minute commerce of prokaryotes, the fungal super-networks under the soil, the slow-grow photosynthesis of plants. We've hardly made any headway into understanding these realms, though they might offer inspiration for all kinds of scientific and technological innovations. And, more importantly, make us reflect on how we understand ourselves.

This last is perhaps why we are so intent on getting into space. Not to look outwards into the void, but to look back at Earth with fresh eyes—even if they are not our own. In 2004 NASA sent two intrepid robots—*Spirit* and *Opportunity*—to Mars, to scout around the Red Planet's alien landscape. They were both set down in Mars' southern hemisphere, and NASA expected contact with the twins to last only about three months, because it was thought that the rough terrain and tumultuous conditions on the planet's surface would surely overcome the rovers in a short time. A robotic "geologist" must be able to move and explore its surroundings, but moving around such a landscape is no simple matter. The rovers looked rather like golf buggies, cruising around while receiving commands sent from Earth over 30 million miles away.

Beyond everyone's wildest hopes, NASA remained in contact with Spirit for six years and with *Opportunity*—or *Oppy*—until February 2019. A mission initially planned for ninety days had lasted fourteen years. After its landing at Meridiani Planum, just south of Mars' equator, *Oppy* drove a total of almost 30 miles, setting an extra-terrestrial driving record. *Oppy* proved to be extremely resilient, dealing with challenges from steep slopes to sand traps; from the hazardous conditions of the Martian winter to blanketing dust storms. Eventually it was a dust storm that proved its demise, blocking the sun for far longer than its solar-powered batteries could endure. The following year, in February 2020, NASA sent another rover, *Perseverance*, to Mars in search of the minute clues that might signify the presence of microbial life from the planet's ancient past. Its progress was tracked by live updates and an online stream that allowed anyone in the world to see through the rover's eyes after it landed gently on Mars' surface. These missions have made some of the first tentative steps to understanding how habitable Mars might be, to know whether extra-terrestrial humans could survive in its arid, russet landscape.

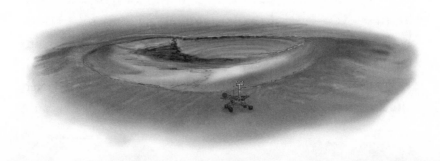

On 3 January 2019, some weeks prior to NASA's final, farewell briefing to *Oppy*, China announced the success of the second phase of its Lunar Exploration project. Named after a lunar deity in Chinese mythology, *Chang'e 4* was a mission to the "dark side" of the moon, the first ever successful "soft landing" in which the craft was not destroyed. On board was a "Lunar Micro Ecosystem," a sealed biosphere containing silkworm and fly eggs and plant seeds from *Arabidopsis thaliana*, potatoes and others. The idea was that these organisms would sustain each other through a reciprocal exchange of carbon dioxide, oxygen and nutrients. And, indeed, the plant seeds germinated, though the experiment was terminated after only two weeks. This was a remarkable milestone: the very first time that insects or plants had grown on the Earth's natural satellite, opening up the possibility for creating sustainable biospheres beyond Earth. The only time seeds had sprouted in space before this had been on the International Space Station, which hangs low in Earth's orbit. It represented a huge jump in the "scientific space race," moving from exploratory geology into the revolutionary possibilities of biology.

The American and Chinese missions are nothing short of incredible, but future missions might yet achieve more. For one, rather than mobile rovers based on animal ideas of moving across the surface, another NASA mission could test out revolutionary plant-inspired models which might have had less trouble overcoming the Martian landscape than those intrepid golf buggies, however ingenious. Major engineering headaches in the design of mobile rovers include how to allow the wheels to endure long enough to cover significant distances, or preventing the vehicle from getting stuck in the sand. One panoramic picture released by NASA became iconic, portraying *Oppy* looking back at its own tracks along a ridge as it headed south. We can clearly see the rim of the crater that the robot has just climbed and feel the palpable sense of relief that everyone at the Jet Propulsion Lab must have felt on seeing that the craft had made it safely there.

Perhaps exploring need not entail mobility though. If we can think differently, we might realise there are radically different solutions to exploring a planet's surface. In June 2017, I received an intriguing email from a major funding body in London. They wanted to explore funding research into "plant intelligence," which might eventually lead to innovations in robotics and AI. The aim was to shift from an animal-centric approach to technological development, to an alternative perspective that might solve problems in new ways. I became part of a research project exploring how plants—their particular ways of growing, moving and interacting with the world—can act as bioinspiration for robotic and AI designs.[1]

What we have found so far has cast doubt on whether there is any need to wheel-drive across the terrain at all. For example, a team of robotics engineers is developing legged "swarm robots" based on the group antics of ants, birds and bees that have much greater flexibility when covering unpredictable ground.[2] Why not go further though? Why do we need to travel across a landscape in search of targets to inspect and sample? Why not *grow* instead? If you are at location A and need to reach location B, one way to do so is simply to grow from A to B. In other words: you can reach B without *leaving* A. This mode of being allows the rover to be in multiple places at once, collecting a network of information just as a plant might. This shift in perspective also completely reconfigures the problem space (and the space problem). Inspiring our rovers with plant-like ways of getting around could allow us not to find a different solution to an existing problem, but rather to redefine the nature of the problem altogether. You don't worry about getting stuck in cracks if you don't have wheels![3]

What we are exploring here is a growing movement to change our perception of plants, something that was also sorely missing in the Chinese mission, as much as it was an impressive feat of bioscience. *Chang'e 4*'s experiment of growing staple organisms under the moon's gravity is pivotal for the success of long-term space missions. The project studies plant growth with the idea that someday astronauts will be able to harvest and cook their own space-grown food. We know well that we couldn't live without plants, so if we are ever to go to space, so must they. Oxygen, food, clothing, medicines or biofuels are essentials for human existence. So it is no coincidence that the seeds chosen for this mission were cotton and potato, together with yeast: the stuff of basic provisions. These crops in their sealed biosphere reveal, though, the rigid way in which these plants are seen. They are understood purely as *resources*. The humans visit space to make new discoveries and push the boundaries of knowledge outwards. Animals travel as key parts of a living system, and perhaps as proxies for ourselves. Plants go to space as mere fodder. Just as we want to explore beyond our own planet, maybe we more urgently need to move past our narrow perspectives. Exploring plants' ways of doing things, seeing into their worlds, potentially as *active* participants in these projects, not just as passive tools, can help us explore worlds beyond Earth.

Plant actors

Though we might not be very good at understanding the subjective experiences of plants, we are extremely astute when it comes to working out how they can benefit us. We are experts at exploiting them. Without them, human life would be untenable. It is not that we don't pay attention to plants. We have become very good at helping them help us: increasing growth rates, removing competitors, cultivating otherwise unusable land. We have even genetically engineered them to do some of this work for us, giving them the ability to resist herbicides or pests, or grow in ways that make harvesting easier. One group recently worked out how to flick genetic switches to cause plants to grow even when in shade; another identified the light-responsive gene networks that direct plant growth.[4] These insights are interesting on their own, but reports never fail to emphasise how we might use this knowledge to increase crop growth under changing climatic conditions. Understanding plants from our current perspective is a means to exploiting them better: to allow them to continue providing for us as we continue to degrade the biosphere. They are passive resources to be manipulated, tended, and even transplanted to space, for our benefit.

What if we could manage to see them differently? Not as objects, but as actors in the ecological networks of which we are part. Like the locked-in patients who cannot communicate their consciousness unless we find ingenious ways to detect it, plants are invisible agents in the ecosystem. They operate at a fundamental level, too slow for us to see and yet vital to our existence. And this relationship is one we are putting a great deal of strain on with our plant-blindness. We will still, for example, happily cut down a hundred ancient oaks to rebuild Notre Dame, sacrificing the living future of silent behemoths for the sake of repairing a relic of human history.

In 2021, I attended Al Gore's talk "The Case for Climate Optimism" at the Frontiers Forum. His presentation was to be followed by a discussion with worldwide experts on climate change and how to ameliorate it.[5] Gore's optimistic argument was that "there is a switch we can flip" to work differently and save the planet, if we can use "science to drive action." On the one hand, this is an attractive idea. Climate activists urge us to support the planting of new forests to replace those that have been decimated in order to suck up carbon from the atmosphere, or to turn our cameras off during Zoom meetings to lower the vast global cost of data storage.[6] These are apparently science-based solutions to a global problem: belief in their efficacy will no doubt make people more compliant. But on the other hand, when Gore expounded to the audience that "the reality we now face implores us to act," I wanted to interject to ask: "act how?" So many of the practical solutions, such as planting trees, seem to be ineffectual sticking plasters on a vast and much deeper problem. We might like to think we can plant our way out of climate change, replacing mature forests with flatpack carbon sinks of fast-growing timber, but the evidence suggests that the two are very much not equivalent, as good as the numbers might look in policy documents.[7]

In the discussions following Gore's talk, I made the argument that revolutionary changes in scientific thinking and practice—which might filter through to genuinely effective practical global responses—require shared frameworks that would allow different disciplines to work together effectively. Such shared perspectives are sorely lacking at the moment. Different specialities are stuck in their own narrow lanes, blinkered to the possibilities of others. One of the first obstacles we *must* face, a key "switch" we need to flick, is to shift our mindset so that we no longer see plants merely as resources for carbon capture or safeguarding food production, but as actors *alongside* us in the climate crisis. We can understand the biology of plants all we like, but if we continue to see them only as the green backdrop to our animal drama on an abiotic stage, we will not solve the problems that we are facing. We are now driving a rapidly escalating shift in the Earth's climate. Before us, the only other multicellular organisms to create such a dramatic change were plants that took over the terrestrial landscape hundreds of millions of years ago. They transformed Earth's atmosphere when they began to photosynthesise and trapped carbon dioxide in their tissues, while pumping out oxygen. We certainly can't undo the damaging changes we have wrought on the climate and biosphere without plants' help, but we need to learn to see them differently for the partnership to work.

This sentiment is echoed in Michael Moore's controversial 2019 documentary *The Planet of the Humans.*[8] He concludes: "I truly believe that the path to change comes from awareness. And that awareness alone can begin to create the transformation. There is a way out of this. We humans must accept that infinite growth on a finite planet is suicide." He makes a plea that "let's for God's sake use scientific knowledge, rather than *something else.*" This awareness, I would argue, must include acknowledging the agency of photosynthetic organisms. Yes, they act as the foundation of the biosphere, link our energy economy with the solar source, but they are not just that. They are also keenly aware of the world they actively shape. It might be that in order to create a sustainable future we have to return to an awareness of the organisms around us, regain our connection with plants as potent co-inhabitants of the planet.

Some plant scientists have begun to try and shake up our views of plants in the global ecosystem. Usually, plants are described in climate change models simply as "passive carbon-fixing entities." My friends and colleagues František Baluška and Stefano Mancuso argue that they "possess a plant-specific intelligence, with which they manipulate both their abiotic and biotic environment, including climate patterns and whole ecosystems." Plants are not ecosystem functions, they argue. Plants and their root networks of symbionts are proactive engineers of their environments that we need to work *with* if we are to undo the changes we have wrought. I would go still further: plants are not just a complex "living air-conditioning system." If we can see plants as *cognitive* beings, we might be able to shift our own perspective on humanity's role in the Earth's biosphere and facilitate plants in rebalancing our own effects on the ecosystem.[9] We could think how they might experience and explore an extra-terrestrial environment, how they might shape it into something habitable for themselves, rather than confining them to encapsulated biotic loops with silkworms, or developing inept, legged robots to tend to them in extra-terrestrial plantations.

Growbots[10]

It might not seem that we will find the kinds of plant-aligned perspectives I am proposing in cutting-edge technology fields such as robotics. These perspectives are apparently more akin to the kind of neo-pagan circles that like to hug trees and consider homeopathy a viable medical alternative. But I want to take a dive into recent technological developments that have germinated from a deeper kind of understanding of plants. There is nothing ephemeral about truly understanding and learning from the radically different way in which plants operate. It is fundamentally world-expanding, and grounded in hard science. It demonstrates how such a perspective shift could have dramatic and concrete global effects.

Robotics has for a long time been zoocentric: making machine animals with metal carapaces and awkward hydraulic joints. They have become ever better at adapting to the environment, tackling the unexpected and avoiding trip-ups with self-righting mechanisms and modularity. We have mimicked in robotic doppelgängers the biomechanics of wall-scaling gecko feet, avian aerodynamics and mammalian gaits, thanks to the investment of vast intellectual and technological resources.[11] But the inherent rigidity and need to *move* as a unit through the environment mean that these robots keep encountering the same problems, as NASA technicians are well aware. MIT's ingenious robot "cheetah" might be able to do a backflip, but if it encounters terrain that is uneven and difficult to traverse, it is stuck.[12] Working on these obstacles within the parameters of traditional robot design will likely always have limitations. Instead, a relatively new area in robotics is creating revolutionary solutions to these problems. Soft robotics takes very different models as its muses. Rather than making metal vertebrates, it uses the Houdini-like flexibility of soft organic structures: octopuses, elephant trunks and earthworms. They give us hydraulic grasping tendrils and pneumatically controlled soft-bodied forms which can move in ways that metallic structures are unable to. They are exceedingly adaptable because of this flexibility, which makes them useful in new ways. They not only offer new solutions to old robotics problems, they make things possible that were previously unthinkable.[13]

Yet the mere change from hard to soft is not sufficient to realise the truly revolutionary aspects of new robotic models. It's not just the expanded biomechanical palette that we can benefit from. Some of these soft models are incredibly tricky to imitate, even with the array of high-tech materials we have at our disposal. Simple motions that rigid robots are incapable of are possible with soft robots, but they can require many different controls and fine, technically difficult coordination. Artificially mimicking the movements of soft animal limbs—the plastic strength of an elephant trunk or the fluid dexterity of an octopus tentacle—is surprisingly hard. An elephant's trunk has nearly 40,000 muscles, all coordinated together. But there are other organisms that lend themselves to soft robotics of a different kind, and have an entirely alternative mode of being which opens up yet more possibilities, without such complex technical challenges. These are, of course, plants: they have both fluidity in their growth and rigidity in form, which is a different proposition from a continually mobile, but uniformly sized robotic limb.

The way that plants move and traverse their surroundings had drawn the curiosity of scientists for many centuries, but Charles Darwin was, as ever, the most eloquent of observers. He was particularly fascinated by the coiling tendrils of cucumbers, which he described as soft springs. He even devoted the bulk of a long monograph to his thoughts on them, though he could not access their inner workings in the depth we are now able to. Researchers have recently identified the fibrous specialised cells that form a strong ribbon through the centre of the cucumber's spring, which draws it into a tight coil without twisting the stem.[14] This kind of mechanism, along with other inspirational plant structures, such as vine hooks for attaching to surfaces, offers a whole new repertoire of mechanical tricks. A key model used in 2020 by researchers at the University of Georgia, much to my delight, was the pole bean. They have developed a robot tendril that is capable of gently twining around and grasping an object just a millimetre in diameter using only a single pneumatic control. It can operate in confined spaces and its motion is monitored by a fibre-optic cable threaded through its core. This spiralling silicone tendril might have applications in all sorts of fields, from sorting through delicate agricultural produce to minute biomedical operations.[15]

We might go even further if we look beyond mere biomimesis, copying the mechanics of how plants do things and the kinds of materials used to do them, using plants as inspiration in the same way we do animals. If we take a step back and think about how plants *use* their tendrils, their behaviours rather than their physical properties, we can play a whole new robotic game. Growing through space rather than moving through it, climbing and grasping rather than locomoting, distributing control and processing through the entire structure, offer a new realm of opportunities. We can design robots that solve problems with diffused intelligence and adaptive forms.[16]

Recently a "growbot" was developed by researchers at Stanford University and University of California, Santa Barbara, which

everts a core of pneumatic plastic tubing to move by growing forward, like the growing tip of a plant shoot continually
producing more cells.[17] It doesn't suffer from the challenges of
friction, uneven surfaces or constrained spaces, it simply everts
over them. It can push through two sticky sheets of fly paper,
or shoot through a bath of glue, by virtue of the outer surface
remaining static. It can be as wide or as narrow as it needs to be,
pushing through small gaps with the sheer force of air pressure.
It can be pierced by sharp objects and remain inflated. It can
grow in any direction, horizontally or vertically, controlled by
specific pneumatic shifts that mimic the differential elongation
of cells which allow plant stems to change direction. It can grow
around a corner or up a wall, and form a rigid hook that can turn
a handle. And it can send sensors to specific locations through its
air-buffered core or use a camera at its tip to direct its movement.
The initial reaction when seeing videos of the growbot is to think
of it as a giant and very purposeful balloon. But the more sober
impression is of a synthetic root, its translucent, inflated form
moving through space and yet static. One thing it can do, which
a root cannot, is to beat a retreat, "un-growing" back through
space by inverting into its core.

It tells us a lot that as technology gets softer, it becomes
more adaptable and attuned to its surroundings, while being
extraordinarily effective. NASA might have the opportunity to
explore Mars with something like a growbot in the future, which
could yield a very different perspective on the planet's surface.
Growbots could still be used to investigate all sorts of impass-
able places, though, until there is another Mars mission: from
brain ventricles and narrow architectural spaces to the deep sea.
Like plants, they can grow adaptively, so they are consummate
generalists. The question is: if we manage to create a robot that
operates as a plant would with incredible efficacy, will it help us
to shift our frameworks and enable us to see how plants can be

allies, not just subjects, in the way we go forward to shape the biosphere? If we can imbue a plantlike robot with human aims and rapid, diffused agency, might it help us to see the intentions and actions of plants in a new light?

Ecological crisis and dignity

In October 2016 I made my way to the Art, Nature and Technology workshop at the Giardino di Daniel Spoerri in Tuscany. The workshop focused on plant behaviour, so Stefano Mancuso and I planned to walk those attending through the wonders of the plant world.[18] The astonishing settings, along with the sumptuous Tuscan food and wine, eased us into some laidback discussions. These quickly started to centre around the appreciation of plant life—including the greens that were just about to be served. As the conversation developed, I had a strong sense of déjà vu: three months earlier, I had been invited to have "a conversation" at the Approaching Plant Consciousness workshop held in the charming Prinzessinnengärten in Berlin. My talk raised ethical questions. I was not surprised, though I didn't have the answers myself. For many years, the implications of the proposition that sentience might extend well beyond the animal world had not troubled me. But now, strolling through the beautiful gardens in Tuscany, I felt the urge to get to the bottom of the ethical questions we had to face if we delved into the minds of plants.

On my flight to Tuscany from Edinburgh, I'd had to change planes in Amsterdam. Dutch Royal Airlines, I soon discovered, took the issue of animal welfare seriously. We were given "ethical" in-flight meals. The bread was from organic grain, processed in local windmills.* The cheese, always a necessity for the Dutch, was produced using sustainable palm oil. The hens that had provided the eggs were stock that lived a world away from the birds subjected to hyper-industrialised production lines for mass consumption. Rondeel, the producer, ensures that they have plenty of room and open air, with different zones for feeding, nesting and laying, and a separate area where they can perch comfortably. They even have 24-hour live webcams to allow egg-eaters to watch how the chickens are doing on the farm.[19] As I read the stories of where each element of my delicious onboard sandwich had come from, I was genuinely impressed at how KLM seemed to have covered all bases when it came to meat, eggs and dairy. Minimising the side effects of globalised transport is also high on KLM's list.

Caring, even if nominally, about the rights of animals and the environmental sustainability of our consumption is not difficult these days. We talk about animal welfare and animal rights because we're concerned about the amount of cruelty shown towards them. In the case of chickens, unless we decide to bury our heads in the sand, we certainly think they feel pain. Which is why Rondeel cares about providing chickens with all the things that make chicken lives worth living. But as we have amply seen, we have a vast blind spot when it comes to caring about plants beyond enhancing their capacities to provide for us. As a passenger with KLM, you applaud their efforts to provide sustainable and well cared-for meat and dairy. But have KLM thought of the garnish, the carrots, peas and potatoes that go with the chicken breast? If the main thesis of this book is correct, plants are intelligent. They have subjective experiences of the world. So shouldn't we care about plants for *their own sake*?

We don't seem to be quite ready to confront the welfare and

---

* Provided by QiZiNi, a food company based in Holland.

rights of plants. Plant life is simply not in the same genre as animal life. In fact, if we cared even a little for the unnecessary stress we inflict on plants, we would have set up ethical committees in research institutions by now, of the very same sort we customarily rely on for the purpose of animal experimentation. Considering this, a memory surfaced of a research visit I paid to Monica Gagliano in Perth, Australia, in December 2013. At the time, it was my understanding that Monica belonged to a plant ethics committee at the University of Western Australia. When I emailed her recently to gather the details for this book, she dispelled my optimistic delusions. My memory had betrayed me. There is no plant ethics committee in Australia or almost anywhere else. It turned out that Monica was in fact overseeing animal research as a member of an *animal* ethics committee, due to her background in coral reef fish ecology. She assured me, "There are no regulations that bind/limit/control plant research from an ethical/welfare/moral standing of plants, so there is no committee overseeing *plant* research."

We are already mulling over the potential future ethical questions surrounding artificial intelligence. Lengthy, erudite discussions are being had on whether machines with cognitive capacities would need to be included in our ethical system; whether we need to limit their scope to prevent them from becoming like us; or how we need to rethink our attitudes to technology that we use if it has a higher degree of awareness. This might be a fascinating intellectual exercise, but it is also true that AI seems to approximate human sapience. It mimics the outputs of human cognition through means that don't really mirror the internal processes of the human mind, at least not yet. But because of this uncanny similarity, we feel duty bound to open the philosophical and political floor to discussing these questions. As the technology develops further over the century and the abilities of this computational intelligence become eerily more similar to ours, these questions will become more pressing.

Plants share far more with us than AI does: they are carbon-based life, with similar metabolic processes and cell structures. We even share ancestry, springing from the same single-celled progenitor that lived billions of years ago. But their sapience, such as it is, is quite different from ours. It is not readily accessible to us, so it does not seem an obvious philosophical move to question the ethical implications of plant consciousness. Most philosophers who have begun to discuss the issue dismiss the possibility that plants are sentient at the outset of their argument, and follow it up with a fatal blow negating the need for an ethical standpoint on plant sentience, arguing that the conclusion of such an argument would lead to absurd implications, even were plants conscious.[20] But the inconvenience and apparent "absurdity" of a line of questioning is not a valid argument against it. We have to face many "inconvenient truths" in the process of undoing, or even slowing, our decimation of the biosphere. Our treatment of plants for their own sake, beyond purely ecological issues, may well be another one. The parallels that are emerging between the ways that plants sense, understand and respond to their environments and the ways that animals do, are making it increasingly difficult to avoid these questions. In fact, our success at tackling the ecological crisis may depend upon facing these issues.[21]

Mahatma Gandhi once said that "the greatness of a nation and its moral progress can be judged by the way its animals are treated." Albert Einstein likewise remarked, "If a man aspires towards a righteous life, his first act of abstinence is from injury to animals." They might have included plants, had they known what they are capable of. Of course, Darwin wrote extensively on how considering the minds of "lower animals" could open up our awareness of how our own minds worked, and our moral sense.[22] Far earlier thinkers than these considered the ethics of plant consumption, from Aristotle's successor Theophrastus to

Pythagoras and Plato, who discussed the relative similarities and differences between plants and animals and their various positions on the moral status this should afford plants.[23] But we stopped asking those questions for a very long time. If plants are intelligent and aware of their surroundings, we cannot turn a blind eye to ethical considerations. So we will end *Planta Sapiens* with what is probably the most difficult question to tackle. Were plants to be given the status of "sentient," would this give them rights that might encumber our exploitation of them? If we allowed them the status of sentient, ethical entities, might we not be able to improve plants' welfare with a little consideration? And shouldn't we?

We have been slow to consider these issues. Though one might have expected them to be at the forefront of this thinking, the Plant Science Research Network's *Decadal Vision for 2020–2030* took a deeply pragmatic stance, considering how best to utilise plants for food security and environmental protection.[24] In stark contrast, plant treatment was in fact discussed in the Federal Ethics Committee on Non-Human Biotechnology (ECNH) constituted by the Swiss Executive Federal Council in 2008. The result was a declaration entitled "The dignity of living beings with regard to plants. Moral consideration of plants for their own sake." ECNH member Florianne Koechlin explained the background of the Swiss declaration in a letter to the editor of The Plant Signaling and Behavior Society's official journal: "The Swiss constitution maintains that the dignity of creatures should be respected. Plants are living beings, so they also have dignity."[25] It takes courage to put "dignity" and "plant" together in one sentence, and yet they did.

In the early twentieth century, Sir J. C. Bose did something many would consider madness today. When taking charge of tree transplantations, he would avoid inflicting unnecessary suffering on the trees by performing anaesthesia on them, even very large individuals. He would create a vast tent to cover the whole tree and filled it with chloroform gas to put the tree to sleep, preventing it being seriously damaged during the moving process. Bose cannot have known the full extent of plant awareness. He knew only that plants could be put to sleep, that they lost awareness of some kind in a way that animals also did when under chloroform. And that was enough for him to take some effort to reduce the stress they experienced when being uprooted and transported to a new spot. Likewise, we needn't have a fully developed understanding of plant consciousness to ask these questions. If we ourselves are creatures that care about the suffering of others, then surely we should consider any organisms that we know can suffer. We need only keep the possibility of plant sentience in our minds, and try to work towards seeing plants in a new way. It can only benefit us all in the long term.

THE HIPPOCAMPUS-
FATTENING FARM

Charles Darwin lived deeply entwined with the branching flow
of ancestry that connects all organisms. Near his death, he
expressed his wish to continue his entanglement with the tree
of life: to be buried under the ancient yew that grew next to St
Mary's Church in Downe. His older brother Erasmus had already
been laid to rest in the churchyard there, and several more of the
family would go to join him eventually. Though the papers published the news that Darwin would have his final wish honoured
at his death, he was not to be allowed to slip quietly into eter-
nity under a tree. His body was instead brought to Westminster
Abbey, where he was given a state funeral full of all the pomp
and circumstance befitting a national treasure.[1] It *looked* like the
right kind of commemoration. But in actual fact, the concerns of
politics and appearances had denied Darwin his peaceful eternal
picnic under the yew.

The powers that be might have listened to the child within Darwin, who knew that the yew was the true tree of the Knowledge of Good and Evil. The church might have seemed like the spiritual focal point of Downe, but it was really the yew, which had stood there long before the chapel was built, an ancient symbol of rebirth and resurrection. Darwin did not wish to be entombed in the rigid halls of history, but to be part of the dynamic flow of life, the endless life cycle of "forms most beautiful and most wonderful." He had devoted himself to bringing a new vision of the natural world to fruition, but was trapped in death by the rigid grasp of tradition. So many of Darwin's ideas have proved to be prescient, despite the fact that he had no access to genetic studies or the kinds of data analyses which are now the bread and butter of biology. He learned to think differently and clearly outside the frameworks in which most of his contemporaries happily confined themselves. He was incredibly erudite, but at the same time, looked with naïve eyes at the world, and so was able to see it in a new way.

We might need to take a leaf out of Darwin's book if we are to attempt to change the way we live in the world in any meaningful way, to create the changes we must in order to avoid the likely disastrous outcomes of our current way of living. In 2006, Sir Ken Robinson gave the most-watched TED talk ever made: "Do Schools Kill Creativity?" It has been watched more than 70 million times, by around 380 million people in 160 countries.[2] Ken's point is, "if you're not prepared to be wrong, you'll never come up with anything original." Our educational system is built on a process of giving school children defined quotas of information to absorb and regurgitate over many years, stifling their instincts to explore for themselves, or to think differently. I think of such institutions as "hippocampus-fattening farms," the hippocampus being a key seat of memory in the brain. Over-education, the hyper-entrainment of young minds into well-trodden paths, like so many vines tethered tightly to trellises, is robbing us of our creative abilities. We might do better being encouraged to *know* less and instead to *think* more.

The biologist Sydney Brenner, who won the Nobel Prize in Physiology, described the power of ignorance in his book, *My Life in Science*. In one interview he said, "I'm a great believer in the power of ignorance. I think you can always know too much." He describes how "one of the things of being an experienced scientist in a subject is that it curtails creativity, because you know too much and you know what won't work." Brenner's argument is that being an "expert" in a field can limit us to only being able to think in one way, cutting off other avenues of exploration. People with different backgrounds, training and perspectives can often breathe new life into well-worn problems, coming up with new solutions that the people immersed firmly in the field would never have allowed themselves to consider.

Science creates its own ivory towers, shutting off access to ideas and problems to anyone outside of scientific circles. Taken as an entity, science assumes that "lay people" will have nothing to add to conversations that touch on scientific matters. It tends to rebuff input from sources external to science, whether they be political, creative, or even commonsensical. But the people who practise science are not just "scientists," they are humans, with all of their ineptitudes, concerns, relationships and creativity. Science is not a self-contained bubble, it is a human endeavour within the rich fabric of human experience. It is inescapably flawed but also full of endless possibility. We just need to allow it to communicate with the rest of the human world, to draw on other threads, other ways of thinking, the ideas of people with expertise in radically different disciplines which might enrich it. This is especially crucial if we are to make science integral to the ways in which humans, scientists and non-scientists alike, are to tackle the problems of the future.

We might recall the invocation of Nobel laureate X-ray crystallographer Richard Axel to thinking "outside the box, between the lines and beyond the horizon" in order to break out of the shell bounding our current frames of thought. Perhaps, if we are ever to come to a new way of thinking, a new way of existing, we have to change our priorities. We have to allow new questions to be asked, to imagine ourselves into the alternate natures of the beings that exist alongside us. If we can *truly* understand what it is like to be a plant, we will learn much about what it means to be human, and how we might be ourselves in ways that work *with* the organic world rather than destroying it. We might uncover a wish, like Darwin, to reconnect with the Tree of Knowledge which roots all life on Earth—to draw on the sapience of plants in order to better comprehend the nature of our own minds.

ACKNOWLEDGEMENTS

This book, like a great many others, is the result of a rich and vibrant ecosystem of interactions, discussions, insights, experiences—and hard work from many quarters. I would like, first of all, to offer my warmest thanks to my agent, Jessica Woollard, who believed so strongly in this project, and eventually had the flash of brilliant insight which allowed it to come to fruition. She brought Natalie and me together to write this book, suspecting that we would work well as a team. From the first meeting at the David Higham offices in 2019, it has been a splendid collaboration. Even the difficulties of Covid-19 did not get in the way of the international teamwork. We have spent many happy hours brainstorming together over Skype and Zoom, even as we've been jotting down these last few words.

After this visit to the offices in London, I went for a walk in Kew Gardens with Carlos Magdalena, author of *The Plant Messiah*. He voiced a passing musing from our discussions as he saw me off to the airport at the gates of Kew: "Hmmmm, plants are sapiens." It sparked the idea for the title of this book, and I am grateful for that vital part of this book's creation.

In my academic life, I owe the greatest debt to František Baluška, Stefano Mancuso and Tony Trewavas. I cannot thank the three of you enough for your visionary thought on the fringes, for having the courage to think differently, and for your constant support.

It had never crossed my mind to do the scientific work behind my philosophical ideas myself, only to collaborate with those who had the labs and resources. But when visiting František in Bonn, trying to persuade him to carry out some experiments in his lab, he suddenly said: "Why don't you do them yourself in Murcia?" I flew back to Spain full of excitement at the possibility, and immediately began generating ideas for how to make it happen. František's suggestion was both a challenge and the tipping point that led to the Minimal Intelligence Lab, MINT Lab, at the University of Murcia.

I've always been fortunate to be funded by agencies that trust ideas over the concrete track record of a publication list. At the very beginning of this work, I had more ideas than results. I'm grateful to the many funding agencies that, in one form or another, have supported my research over the last two decades. I'd like to give special thanks to Fundación Séneca, the Agency for Science and Technology of the Region of Murcia (Spain), without whom I wouldn't have been able to found MINT Lab. Much of what I have learned and experienced that generated this book was made possible by a "Stays of professors and senior researchers in foreign centres" fellowship awarded by the Spanish Ministry of Education, Culture and Sport. It provided me with the financial peace of mind that my family and I needed during our stay in Edinburgh. I would like to express my gratitude to my hosts at the University of Edinburgh, Andy Clark and Tony Trewavas, for their constant support (and for sharing their offices with me!).

Thanks too, to the three musketeers, Manuel Heras-Escribano, Vicente Raja and Miguel Segundo-Ortín, predocs who became colleagues. Time flies!

To the MINT Lab team, Jacobo Blancas, Anna Finke, Adrian

Frazier, Jonny Lee and Aditya Ponkshe, and to the many visitors past, present (and hopefully future).

I am so grateful to the co-authors of many of my academic articles on plants. Among them, I would particularly like to mention Charles Abramson, František Baluška, François Bouteau, Karl Friston, Monica Gagliano, Ángel García Rodríguez, Fred Keijzer, Dave Lee, Adam Linson, Stefano Mancuso, Pedro Mediano, Paula Silva, Andrew Sims, Gustavo Maia Souza and Tony Trewavas.

To Almudena Gutiérrez Abbad, director of the Agroforestry Experimentation Service at University of Murcia, and all her team. To Juan Francisco Miñarro Jiménez, at the Mechanical Workshop, and Fernando Ruiz Abellán, at the Electronics Workshop, for the amazing technical support they provide.

To the Faculty and the Department of Philosophy at University of Murcia for providing the most hospitable of working environments anyone could ask for.

I would like to give very special thanks to Liz Van Volkenburgh, for her constant support throughout the years. We met when I first attended the meetings of the Plant Neurobiology Society, and many times subsequently at the Plant Signaling & Behavior Society meetings that she chaired. Her balanced, tempered and wise advice has been invaluable. She read the full manuscript of this book and saved me from making some factual errors. Of course, any mistakes or erroneous ideas that are presented here are not to be seen as in any way her responsibility.

I am very grateful to Tony Trewavas too for his generosity, both intellectual and material. He and his wife Val showed such warmth and kindness to my family and I during our stay with them, and I found myself driving back to Spain with a car boot filled with the plant science books that Tony had generously given me.

I'd like to say something about the detractors of mine and my colleagues' work. Of course, there can be no progress in any field without tension and disagreements. Sometimes these are fierce, I'm afraid. But, despite that, I'm grateful to have had the chance to test my ideas and work in my interactions with Lincoln Taiz, Michael Blatt and David Robinson. Criticism and opposition can only have made me work harder. If they happen to read this book at all, I hope they might consider rethinking some of their criticisms.

To my children, Hortensia and Paquillo. A book project and adolescence under the same roof? No small feat! To my parents who—once upon a time—came back home from a trip with a copy of *Platero y yo* as a present for me, at an age I couldn't yet grasp how important it was. To my sisters Pingo and Maena, and to my handful of real friends. They all know why.

Last, in loving memory of Jim Edwards (1939–2021) and Rosa Alcázar Leante (1961–2019). Jim was my PhD supervisor back in Glasgow in the nineties. I owe him the greatest intellectual debt I can imagine. Rosa was the secretary admin of the Department of Philosophy when I first set my lab up. She will always be the *guardian angel* of MINT Lab.

NOTES

Introduction: Putting Plants to Sleep

1. Eisner, T. (1981), "Leaf folding in a sensitive plant: A defensive thorn-exposure mechanism?" *Proceedings of the National Academy of Sciences* 78: 402–404.
2. Hedrich, R., Neher, E. (2018), "Venus flytrap: How an excitable, carnivorous plant works," *Trends in Plant Science* 23: 220–234.
3. Yokawa, K., Kagenishi, T., Pavlovic, A., Gall, S., Weiland, M., Mancuso, S., Baluška, F. (2018), "Anaesthetics stop diverse plant organ movements, affect endocytic vesicle recycling and ROS homeostasis, and block action potentials in Venus flytraps," *Annals of Botany* 122: 747–756.
4. Bouteau, F., Grésillon, E., Chartier, D., Arbelet-Bonnin, D., Kawano, T., Baluška, F., Mancuso, S., Calvo, P., Laurenti, P. (2021), "Our sisters the plants? Notes from phylogenetics and botany on plant kinship blindness," *Plant Signaling & Behavior* 16: 12, 2004769.
5. After Domains, the groups divide into Kingdoms, Phyla, Classes, Orders, Families, Genera and, finally, Species.
6. The "father of experimental biology," the French physiologist Claude Bernard, said it first: "all life is defined by the susceptibility to anaesthesia." He was also convinced that

the physiological functions of all organisms rely on the same underlying mechanisms and "sensitivity" to the environment. Bernard, C., *Leçons sur les phénomènes de la vie communs aux animaux et aux végétaux*, Lectures on Phenomena of Life Common to Animals and Plants. Paris: Ballliere and Son, 1878. See also Kelz, M. B., Mashour, G. A. (2019), "The biology of general anesthesia from paramecium to primate," *Current Biology* 29: R1199–R1210.

7. Grémiaux, A., Yokawa, K., Mancuso, S., Baluška, F. (2014), "Plant anesthesia supports similarities between animals and plants: Claude Bernard's forgotten studies," *Plant Signaling & Behavior* 9: e27886.

8. Laothawornkitkul, J., Taylor, J. E., Paul, N. D., Hewitt, C. N. (2009), "Biogenic volatile organic compounds in the Earth system," *New Phytologist* 183: 27–51.

9. Tsuchiya, H. (2017), "Anesthetic agents of plant origin: A review of phytochemicals with anesthetic activity," *Molecules* 22: 1369; Baluška, F., Yokawa, K., Mancuso, S., Baverstock, K. (2016), "Understanding of anesthesia—Why consciousness is essential for life and not based on genes," *Communicative & Integrative Biology* 9: e1238118.

10. For a rich and lyrical account of the human relationship with and use of botanicals, see Pollan, M. (2001), *The Botany of Desire: A Plant's-eye View of the World*. London: Random House.

11. For an analysis of snapping durations and speeds under different environmental conditions, see Poppinga, S., Kampowski, T., Metzger, A., Speck, O., Speck, T. (2016), "Comparative kinematical analyses of Venus flytrap (*Dionaea muscipula*) snap traps," *Beilstein Journal of Nanotechnology* 7: 664–674.

12. Silvertown, J., Gordon, D. M. (1989), "A framework for plant behavior," *Annual Review of Ecology and Systematics*

20: 349–366; Karban, R. (2008), "Plant behaviour and communication," *Ecology Letters* 11: 727–739.

13. Cvrčková, F., Žárský, V., Markoš, A. (2016), "Plant studies may lead us to rethink the concept of behavior," *Frontiers in Psychology* 7: 622.

14. Grémiaux, A., Yokawa, K., Mancuso, S., Baluška, F. (2014), "Plant anesthesia supports similarities between animals and plants: Claude Bernard's forgotten studies," *Plant Signaling & Behavior* 9: e27886.

15. Schwartz, A., Koller, D. (1986), "Diurnal phototropism in solar tracking Leaves of *Lavatera cretica*," *Plant Physiology* 80: 778–781.

16. Eelderink-Chen, Z., Bosman, J., Sartor, F., Dodd, A. N., Kovács, Á. T., Merrow, M. (2021), "A circadian clock in a nonphotosynthetic prokaryote, *Science Advances* 7: eabe2086; Cashmore, A. R. (2003), "Cryptochromes: enabling plants and animals to determine circadian time," *Cell* 114: 537–543.

17. Descartes described his theories about the pineal in both the *Treatise of Man* (written before 1637, published posthumously, in Latin (1662) and French (1664), and in his last book, *The Passions of the Soul* (1649).

18. Dubbels, R., Reiter, R. J., Klenke E., Goebel, A., Schnakenberg, E., Ehlers, C., Schiwara, H., Schloot, W. (1995), "Melatonin in edible plants identified by radioimmunoassay and by HPLC-MS," *Journal of Pineal Research* 18: 28–31; Hattori, A., Migitaka, H., Iigo, M., Yamamoto, K., Ohtani-Kaneko, R., Hara, M., Suzuki, T., Reiter, R. J. (1995), "Identification of melatonin in plants and its effects on plasma melatonin levels and binding to melatonin receptors in vertebrates," *Biochemistry and Molecular Biology International* 35: 627–634.

19. Balcerowicz, M., Mahjoub, M., Nguyen, D., Lan, H.,

Stoeckle, D., Conde, S., Jaeger, K. E., Wigge, P. A., Ezer, D. (2021), "An early-morning gene network controlled by phytochromes and cryptochromes regulates photomorphogenesis pathways in Arabidopsis," *Molecular Plant* 14 (6): 983.

20. Calvo, P., Trewavas, A. (2020), "Cognition and intelligence of green plants: Information for animal scientists," *Biochemical and Biophysical Research Communications* 564: 78–85.

21. Mellerowicz, E. J., Immerzeel, P., Hayashi, T. (2008), "Xyloglucan: the molecular muscle of trees," *Annals of Botany* 102: 659–665; Gorshkova, T., Brutch, N., Chabbert, B., Deyholos, M., Hayashi, T., Lev-Yadun, S., Mellerowicz, E. J., Morvan, C., Neutelings, G., Pilate, G. (2012), "Plant fiber formation: state of the art, recent and expected progress, and open questions," *Critical Reviews in Plant Sciences* 31: 201–228.

22. Calvo, P., Gagliano, M., Souza, G. M., Trewavas, A. (2020), "Plants are intelligent, here's how," *Annals of Botany* 125: 11–28.

23. Biernaskie, J. M. (2011), "Evidence for competition and cooperation among climbing plants," *Proceedings of the Royal Society B: Biological Sciences* 278: 1989–1996.

1: Plant Blindness

1. James, W. (1890), *The Principles of Psychology*. London: Macmillan.

2. Bar-On, Y. M., Phillips, R., Milo, R. (2018), "The biomass distribution on Earth," *Proceedings of the National Academy of Sciences* 115: 201711842.

3. Alcaraz Ariza, F. (1998), *Guía de las plantas del Campus Universitario de Espinardo*. EDITUM.

4. Balas, B., Momsen, J. L. (2014), "Attention 'blinks'

differently for plants and animals," *CBE—Life Sciences Education* 13: 437–443; Shapiro, K. L., Arnell, K. M., Raymond, J. E. (1997), "The attentional blink," *Trends in Cognitive Sciences* 1: 291–296.

5. Norretranders, T. (1998), *The User Illusion*. New York: Viking.
6. Wandersee, J. H., Schussler, E. E. (2001), "Towards a theory of plant blindness," *Plant Science Bulletin* 47: 2–9.
7. Wandersee, J. H., Schussler, E. E. (1999), "Preventing plant blindness," *American Biology Teacher* 61: 82–86; Wandersee and Schussler (2001), 6.
8. See Kew, Royal Botanic Gardens' *State of the World's Plants and Fungi* report, available online at www.kew.org/SOTWPF.
9. Richards, D. D., Siegler, R. S. (1984), "The effects of task requirements on children's life judgments," *Child Development* 55: 1687–1696; Richards, D. D., Siegler, R. S. (1986), "Children's understanding of the attributes of life," *Journal of Experimental Child Psychology* 42: 1–22; Bebbington, A. (2005), "The ability of A-level students to name plants," *Journal of Biological Education* 39: 62–67; Yorek, M., Sahin, M., Aydin, H. (2009), "Are animals 'more alive' than plants? Animistic-anthropocentric construction of life concept," *Eurasia Journal of Mathematics, Science & Technology Education* 5: 369–378.
10. Brenner, E. D. (2017), "Smartphones for teaching plant movement," *The American Biology Teacher* 79: 740–745.
11. Lawrence, N., Calvo, P. (2022), "Learning to see 'green' in an ecological crisis." In Weir, L., ed., *Philosophy as Practice in the Ecological Emergency: An Exploration of Urgent Matters*. Berlin: Springer.
12. Lovejoy, A. O. (1936), *The Great Chain of Being: A Study of the History of an Idea*. Cambridge, MA: Harvard University Press.

13. Gibson, J. J. (1979), *The Ecological Approach to Visual Perception*. Boston, MA: Houghton Mifflin.

14. Khattar, J., Calvo, P., Vandebroek, I., Pandolfi, C., Dahdouh-Guebas, F. (2022), "Understanding trans-disciplinary perspectives of plant intelligence: is it a matter of science, language or subjectivity?" *Journal of Ethnobiology and Ethnomedicine* 18: 41.

15. Descola, P. (2009), "Human natures," *Social Anthropology* 17: 145–157; Balding, M., Williams, K. J. H. (2016), "Plant blindness and the implications for plant conservation," *Conservation Biology* 30: 1192–1199.

16. Churchland, P. S. (2002), *Brain-wise: Studies in Neurophilosophy*. Cambridge, MA: MIT Press.

17. Barnes, R. S. K, Hughes, R. N. (1999), *An Introduction to Marine Ecology*, pp. 117–41. 3rd edition, Oxford: Blackwell Science.

18. Fox, M. D., Elliott Smith, E. A., Smith, J. E., Newsome, S. D. (2019), "Trophic plasticity in a common reef-building coral: Insights from δ13 C analysis of essential amino acids," *Functional Ecology* 33: 2203–2214.

19. Churchland, P. S. (1986), *Neurophilosophy: Toward a Unified Science of the Mind-brain*. Cambridge, MA: MIT Press.

20. Qi, Y., Wei, W., Chen, C., Chen, L. (2019), "Plant root-shoot biomass allocation over diverse biomes: A global synthesis," *Global Ecology and Conservation* 18: e00606.

21. Hodge, A. (2009), "Root decisions," *Plant, Cell and Environment* 32: 628–640; Novoplansky, A. (2019), "What plant roots know?" *Seminars in Cell and Developmental Biology* 92: 126–133.

22. Baluška, F., Mancuso, S., Volkmann, D., Barlow, P. W. (2009), "The 'root-brain' hypothesis of Charles and Francis Darwin: Revival after more than 125 years," *Plant Signaling & Behavior* 4: 1121–1127.

23. Baluška, F., Mancuso, S. (2009), "Plants and animals: Convergent evolution in action?" In F. Baluška, ed., *Plant-Environment Interactions: From sensory plant biology to active plant behavior.* Berlin: Springer, pp. 285–301.

24. Barlow, P. W. (2006), "Charles Darwin and the plant root apex: closing a gap in living systems theory as applied to plants." In Baluška, F., Mancuso, S., Volkmann D., eds., *Communication in Plants*, pp. 37–51. Berlin: Springer; Kutschera, U., Nicklas, K. J. (2009), "Evolutionary plant physiology: Charles Darwin's forgotten synthesis," *Naturwissenschaften* 96: 1339–54.

25. Mackay, D. S., Savoy, P. R., Grossiord, C., Tai, X., Pleban, J. R., Wang, D. R., McDowell, N. G., Adams, H. D., Sperry, J. S. (2020). "Conifers depend on established roots during drought: results from a coupled model of carbon allocation and hydraulics," *New Phytologist* 225: 679-692. Mackay quoted in Hsu, C. (2 Jan 2020), "How do conifers survive droughts? Study points to existing roots, not new growth," *UBNow* www.buffalo.edu/ubnow/campus.host.html/content/ shared/university/news/ub-reporter-articles/stories/2020/01/ conifers-drought.detail.html

26. Sheldrake, M. (2020), *Entangled Life: How Fungi Make Our Worlds, Change Our Minds and Shape Our Futures.* London: Random House.

27. Smith, M. L., Bruhn, J. N., Anderson, J. B. (1992), "The fungus *Armillaria bulbosa* is among the largest and oldest living organisms," *Nature* 356: 428–431.

28. Bell, B. F. (1981), "What is a plant? Some children's ideas," *New Zealand Science Teacher* 31: 10–14.

29. Camerarius, R. J. (1694). *De Sexu Plantarum Epistola.* University of Tübingen, Germany. See also Žárský, V., Tupý, J. (1995), "A missed anniversary: 300 years after Rudolf Jacob Camerarius. 'De sexu plantarum epistola,'" *Sexual Plant*

Reproduction 8: 375–376; Funk, H. (2013), "Adam Zalužanský's 'De sexu plantarum' (1592). An early pioneering chapter on plant sexuality," *Archives of Natural History* 40: 244–256.

30. Specht, C. D., Bartlett, M. E. (2009), "Flower evolution: the origin and subsequent diversification of the angiosperm flower," *Annual Review of Ecology, Evolution, and Systematics* 40: 217–243; Doyle, J. A. (2012), "Molecular and fossil evidence on the origin of angiosperms," *Annual Review of Earth and Planetary Sciences* 40: 301–326.

31. Beiler, K. J., Durall, D. M., Simard, S. W., Maxwell, S. A., Kretzer, A. M. (2010), "Mapping the wood-wide web: mycorrhizal networks link multiple Douglas-fir cohorts," *New Phytologist* 185: 543–553.

32. Kull, K. (2016), "The biosemiotic concept of the species," *Biosemiotics* 9: 61–71.

33. Nakagaki, T., Yamada, H., Tóth, Á. (2000), "Maze-solving by an amoeboid organism," *Nature* 407: 470.

34. Sanders, D., Nyberg, E., Eriksen, B., Snæbjørnsdóttir, B. (2015), " 'Plant blindness': Time to find a cure," *The Biologist* 62: 9.

2: Seeking a Plant's Perspective

1. Cited in Desmond, A., Moore, J. R. (1992), *Darwin*. London: Penguin. See also Darwin to J. Hooker, 5 Mar. 1863, Darwin Archive, Cambridge University Library, 115: 184; De Beer, Sir G., "Darwin's Journal," *Bulletin of the British Museum (Natural History)*, Historical Series 2 (1959), 16; Colp, R. (1977), *To Be an Invalid*. Chicago: University of Chicago Press, pp. 74–5; F. Darwin (1887), *Life and Letters of Charles Darwin*, 3 vols; 3: 312–13; Allan, M. (1977), *Darwin and His Flowers: The Key to Natural Selection*. London: Faber & Faber, ch. 12 (from Desmond and Moore).

2. Darwin, C. (1865), "On the movements and habits of

climbing plants," *Botanical Journal of the Linnean Society* 9: 1–118. Darwin had been inspired by the work of the American botanist Asa Gray at Harvard. In his essay on climbing plants, Darwin (1865, p. 1) wrote, "I was led to this subject by an interesting, but too short, paper by Professor Asa Gray (1858) on the movements of the tendrils of some Cucurbitaceous plants. He sent me seeds, and on raising some plants I was so much fascinated and perplexed by the revolving movements of the tendrils and stems, which movements are really very simple, though appearing at first very complex, that I procured various other kinds of Climbing Plants, and studied the whole subject." (from Isnard, S., Silk, W. K. (2009), "Moving with climbing plants from Charles Darwin's time into the 21st century," *American Journal of Botany* 96: 1205–1221).

3. Desmond and Moore, *Darwin*, 42–3. Our interpretation of these "springs" today owes its origins to Darwin's work (Gerbode, S. J., Puzey, J. R., McCormick, A. G., Mahadevan, L. (2012), "How the cucumber tendril coils and overwinds," *Science* 337: 1087).

4. Darwin, C. (1875), *The Movements and Habits of Climbing Plants*, pp. 12–13. London: John Murray. For a retrospective view, see Heslop-Harrison, J. (1979), "Darwin and the movement of plants: A retrospect." In *Proceedings of the 10th International Conference on Plant Growth Substances*, Madison, Wisconsin, 22–26 July 1979, pp. 3–14. Berlin and Heidelberg: Springer. (This text corresponds to a lecture given for the centenary of the publication of Darwin's *The Power of Movement in Plants*, 1880.)

5. De Chadarevian, S. (1996), "Laboratory science versus country-house experiments. The controversy between Julius Sachs and Charles Darwin," *British Journal for the History of Science* 29: 17–41. See also Calvo, P., Trewavas, A. (2020),

"Physiology and the (neuro)biology of plant behaviour: A farewell to arms," *Trends in Plant Science* 25: 214–216.

6. Hutton was the founder of modern geology. See Hutton, J. (1788), "Theory of the Earth; or an investigation of the laws observable in the composition, dissolution, and restoration of land upon the Globe," *Transactions of the Royal Society of Edinburgh*, vol. 1, Part 2, pp. 209–304.

7. See Charles Darwin's *Notebook B*, 1837, stored in Cambridge University Library, for his earliest Tree of Life sketch.

8. Burnett, F. H. (1911), *The Secret Garden*. New York: Frederick A. Stokes.

9. Dawkins, R. (1996), *Climbing Mount Improbable*. New York: Norton.

10. Land, M. F., Fernald, R. D. (1992), "The evolution of eyes," *Annual Review of Neuroscience* 15: 1–29.

11. Heslop-Harrison, "Darwin and the movement of plants: A retrospect"; De Chadarevian, "Laboratory science versus country-house experiments."

12. Calvo, P., Baluška, F., Trewavas, A. (2021), "Integrated information as a possible basis for plant consciousness," *Biochemical and Biophysical Research Communications.* 564: 158–165.

13. Pirici, A., Calvo, P. (2022), "Sensing the living: promoting the perception of plants," *Cluj Cultural Centre–Studiotopia– Art meets Science in the Anthropocene.*

14. https://youtu.be/FtCFCkQsBtg. Thanks to Stefano Mancuso for first bringing this clip to my attention.

15. Ebel, F., Hagen, A., Puppe, K., Roth, H. J., Roth, J. (1974), "Beobachtungen über das bewegungsverhalten des Pollinariums von *Catasetum Jimbriatum* Lindl. während Abschuß, Flug und Landung," *Flora* 163: 342–356; Nicholson, C.C., Bales, J. W., Palmer-Fortune, J. E.,

Nicholson, R. G. (2008), "Darwin's bee-trap: The kinetics of Catasetum, a new world orchid," *Plant Signaling & Behavior* 3: 19–23; Simons, P. (1992), *The Action Plant, Movement and Nervous Behavior in Plants*. Oxford: Blackwell.

16. Darwin, C. (1962), "Catasetidæ, the most remarkable of all orchids." In *On the Various Contrivances by which British and Foreign Orchids are Fertilised by Insects*. London: John Murray, pp. 211–85. See also Darwin, C. (1862), "On the three remarkable sexual forms of *Catasetum tridentatum*, an orchid in the possession of the Linnean Society," *Proceedings of the Linnean Society of London* (Botany) 6: 151–157; Darwin, C. (1877), *The Different Forms of Flowers on Plants of the Same Species*. London: John Murray; Darwin, C. (1876), *The Effects of Cross and Self Fertilisation in the Vegetable Kingdom*. London: John Murray.

17. Heider, F., Simmel, M. (1944), "An experimental study of apparent behavior," *American Journal of Psychology* 57: 243–249. See https://www.youtube.com/watch?v=VTNmLt7QX8E

18. Scholl, B. J., Tremoulet. P. D. (2000), "Perceptual causality and animacy," *Trends in Cognitive Sciences* 4: 299–309.

19. Agassi, J. (1964), "Analogies as generalizations," *Philosophy of Science* 31: 4; Agassi, J., "Anthropomorphism in science." In Wiener, P. P., ed. (1968, 1973), *Dictionary of the History of Ideas: Studies of Selected Pivotal Ideas*, pp. 87–91. New York: Scribner.

20. Reed, E. S. (2008), *From Soul to Mind: The Emergence of Psychology from Erasmus Darwin to William James*. New Haven and London: Yale University Press.

21. Andrews, K., Huss, B. (2014), "Anthropomorphism, anthropectomy, and the null hypothesis," *Biology & Philosophy* 29: 711–729.

22. Taiz, L., Alkon, D., Draguhn, A., Murphy, A., Blatt, M.,

Hawes, C., Thiel, G., Robinson, D. G. (2019), "Plants neither possess nor require consciousness," *Trends in Plant Science* 24: 677–687.

23. Calvo and Trewavas, "Physiology and the (neuro)biology of plant behaviour."

24. Raja, V., Silva, P. L., Holghoomi, R., Calvo, P. (2020), "The dynamics of plant nutation," *Scientific Reports* 10: 19465.

25. Calvo, P. (2016), "The philosophy of plant neurobiology: A manifesto," *Synthese* 193: 1323–1343.

3: Smart Plant Behaviour

1. http://www.linv.org

2. Mugnai, S., Azzarello, E., Masi, E., Pandolfi, C., Mancuso, S. (2015), "Nutation in plants." In Mancuso, S., Shabala, S., eds., *Rhythms in Plants*, pp. 19–34. Berlin: Springer.

3. Darwin, C., Darwin, F. (1880), *The Power of Movement in Plants*. London: John Murray.

4. Baillaud, L. (1962), "Les mouvements d'exploration et d'enroulement des plantes volubiles." In Aletse, L. et al., eds., *Handbuch der Pflanzenphysiologie*, pp. 637–715. Berlin: Springer; Millet, B., Melin, D., Badot, P.-M. (1988), "Circumnutation in *Phaseolus vulgaris*. I. Growth, osmotic potential and cellular structure in the free-moving part of the shoot," *Physiologia Plantarum* 72: 133–138; Badot, P.-M., Melin, D., Garrec, J. P. (1990), "Circumnutation in *Phaseolus vulgaris* L. II. Potassium content in the free-moving part of the shoot," *Plant Physiology and Biochemistry* 28: 123–130; Millet, B., Badot, P.-M. (1996), "The revolving movement mechanism in *Phaseolus*; New approaches to old questions." In Greppin, H., Degli Agosti, R., Bonzon, M., eds., *Vistas on Biorhythmicity*, pp. 77–98. Geneva: University of Geneva; Caré, A. F., Nefedev, L., Bonnet, B., Millet, B., Badot, P.-M. (1998), "Cell elongation

and revolving movement in *Phaseolus vulgaris* L. twining shoots," *Plant and Cell Physiology* 39: 914–921.

5. Darwin, C. (1875), *The Movements and Habits of Climbing Plants*, pp. 12–13. London: John Murray.

6. Desmond, A., Moore, J. R. (1992), *Darwin*. London: Penguin.

7. With the assistance of Vicente Raja—a former student, currently a postdoc at the Brain and Mind Institute at Western University in Canada, and, at the time, a visiting researcher at MINT Lab. Details of the experimental setting and videos can be found in Calvo, P., Raja, V., Lee., D. N. (2017), "Guidance of circumnutation of climbing bean stems: An ecological exploration," *bioRxiv* 122358; Raja, V., Silva, P. L., Holghoomi, R., Calvo, P. (2020), "The dynamics of plant nutation," *Scientific Reports* 10: 19465.

8. Maria Stolarz, from the Institute of Biology and Biochemistry at Maria Curie-Skłodowska University in Lublin, Poland, was kind enough to graph the movement for us with her program, Circumnutation Tracker, the first free and open-source tool for the analysis of plant movements of revolution. See Stolarz, M., Żuk, M., Król, E., Dziubińska, H. (2014), "Circumnutation Tracker: novel software for investigation of circumnutation," *Plant Methods* 10: 24.

9. Segundo-Ortin, M., Calvo, P. (2019), "Are plants cognitive? A reply to Adams," *Studies in History and Philosophy of Science* 73: 64–71.

10. Kumar, A., Memo, M., Mastinu, A. (2020), "Plant behaviour: an evolutionary response to the environment?" *Plant Biology* 22: 961–970.

11. Vandenbussche, F., Van Der Straeten, D. (2007), "One for all and all for one: Cross-talk of multiple signals controlling the plant phenotype," *Journal of Plant Growth Regulation* 26: 178–187; Hou, S., Thiergart, T., Vannier, N., Mesny, F.,

Ziegler, J., Pickel, B., Hacquard, S. (2021), "A microbiota–root–shoot circuit favours *Arabidopsis* growth over defence under suboptimal light," *Nature Plants* 7: 1078–1092.

12. For the wider context to Darwin's quote in relation to the "root-brain" hypothesis, see Baluška, F., Mancuso, S., Volkmann, D., Barlow, P. (2009), "The 'root-brain' hypothesis of Charles and Francis Darwin: Revival after more than 125 years," *Plant Signaling & Behavior* 4: 1121–1127.

13. Allen, P. H. (1977), *The Rain Forests of Golfo Dulce*. Stanford, CA: Stanford UP.

14. Bodley, J. H., Benson, F. C. (1980), "Stilt-root walking by an iriateoid palm in the Peruvian Amazon," *Biotropica* 12: 67–71.

15. Leopold, A. C., Jaffe, M. J., Brokaw, C. J., Goebel, G. (2000), "Many modes of movement," *Science* 288: 2131–2132; Huey, R. B., Carlson, M., Crozier, L., Frazier, M., Hamilton, H., Harley, C., Kingsolver, J. G. (2002), "Plants versus animals: do they deal with stress in different ways?" *Integrative and Comparative Biology* 42: 415–423.

16. Suetsugu, K., Tsukaya, H., Ohashi, H. (2016), "*Sciaphila yakushimensis* (Triuridaceae), A new mycoheterotrophic plant from Yakushima Island, Japan," *Journal of Japanese Botany* 91: 1–6.

17. Baldwin, I. T., Halitschke, R., Paschold, A., von Dahl, C. C., Preston, C. A. (2006), "Volatile signaling in plant-plant interactions: 'talking trees' in the genomics era," *Science*, 311(5762): 812–815; Dicke, M., Agrawal, A. A., Bruin, J. (2003), "Plants talk, but are they deaf?" *Trends in Plant Science*, 8(9): 403–405.

18. Orrock, J., Connolly, B., Kitchen, A. (2017), "Induced defences in plants reduce herbivory by increasing cannibalism," *Nature Ecology & Evolution* 1: 1205–1207.

19. Ryan, C. M., Williams, M., Grace, J., Woollen, E.,

Lehmann, C. E. R. (2017), "Pre-rain green-up is ubiquitous across southern tropical Africa: implications for
temporal niche separation and model representation," *New
Phytologist* 213: 625–633.

20. Atamian, H. S., Creux, N. M., Brown, E. A., Garner, A. G.,
Blackman, B. K., Harmer, S. L. (2016), "Circadian regulation
of sunflower heliotropism, floral orientation, and pollinator
visits," *Science* 353: 587–90.

21. Fisher, F. J. F., Fisher, P. M. (1983), "Differential starch dep-
osition: a 'memory' hypothesis for nocturnal leaf-movements
in the suntracking species *Lavatera cretica L.*," *New
Phytologist* 94: 531–536.

22. A discussion of *Lavatera cretica* and nocturnal reorienta-
tion in the context of the quest for plant cognition can be
found in Calvo Garzón, F. (2007), "The quest for cognition
in plant neurobiology," *Plant Signaling & Behavior* 2: e1.
See also García Rodríguez, A., Calvo Garzón, P. (2010), "Is
cognition a matter of representations? Emulation, teleology,
and time-keeping in biological systems," *Adaptive Behavior*
18: 400–415.

23. Mittelbach, M., Kolbaia, S., Weigend, M., Henning, T.
(2019), "Flowers anticipate revisits of pollinators by learning from previously experienced visitation intervals," *Plant
Signaling & Behavior* 14: 1595320.

24. Novoplansky, A. (2009), "Picking battles wisely: Plant
behaviour under competition," *Plant, Cell & Environment*
32: 726–741.

25. De Kroon, H., Visser, E. J. W., Huber, H., Hutchings, M.
J. (2009), "A modular concept of plant foraging behaviour:
The interplay between local responses and systemic control,"
Plant, Cell & Environment 32: 704–712.

26. Plants, for instance, are able to read their own shapes
(Hamant, O., Moulia, B. (2016), "How do plants read

their own shapes?" *New Phytologist* 212: 333e337); they
can perceive sound (Khaita, T. I., Obolskib, U., Yovelc, Y.,
Hadanya, L. (2019), "Sound perception in plants," *Seminars
in Cell & Developmental Biology* 92: 134–138); sense magnetic fields (Galland, P., Pazur, A. (2005), "Magnetoreception
in plants," *Journal of Plant Research* 118: 371–389; Maffei,
M.E. (2014), "Magnetic field effects on plant growth, development, and evolution," *Frontiers in Plant Science* 5: 445).
Plants are able to interpret many different cues related to
light (Paik, I., Huq, H. (2019), "Plant photoreceptors: Multi-
functional sensory proteins and their signaling T networks,"
Seminars in Cell and Developmental Biology 92: 114–121).
Plants can likewise feel the heat (Vu, L. D., Gevaert, K., De
Smet, I. (2019), "Feeling the heat: Searching for plant ther-
mosensors," *Trends in Plant Science* 24: 210–219). And many
others. For a review, see Calvo, P., Trewavas, A. (2020),
"Cognition and intelligence of green plants: Information for
animal scientists," *Biochemical and Biophysical Research
Communications* 564: 78–85.

27. Dittrich, M., Mueller, H.M., Bauer, H., Peirats-Llobet,
M., Rodriguez, P. L., Geilfus, C.-M., Carpentier, S. C., Al
Rasheid, K. A. S., Kollist, H., Merilo, E., Herrmann, J.,
Müller, T., Ache, P., Hetherington, A., Hedrich, R. (2019),
"The role of Arabidopsis ABA receptors from the PYR/PYL/
RCAR family in stomatal acclimation and closure signal
integration," *Nature Plants* 5: 1002–1011.

28. Xu, B., Long, Y., Feng, X., Zhu, X., Sai, N., Chirkova, L.,
Betts, A., Herrmann, J., Edwards, E. J., Okamoto, M.,
Hedrich, R., Gilliham, M. (2021), "GABA signalling modulates stomatal opening to enhance plant water use efficiency
and drought resilience," *Nature Communications* 12: 1–13.

29. Schenk, H. J., Callaway, R. M., Mahall, B. E. (1999),
"Spatial root segregation: Are plants territorial?" *Advances

in Ecological Research 28: 145–180; Gruntman, M., Novoplansky, A. (2004), "Physiologically-mediated self/nonself discrimination in roots," *Proceedings of the National Academy of Sciences* 101: 3863–3867; Falik, O., Reides, P., Gersani, M., Novoplansky, A. (2005), "Root navigation by self inhibition," *Plant, Cell & Environment* 28: 562–569; Novoplansky, A. (2019), "What plant roots know?" *Seminars in Cell and Developmental Biology* 92: 126–133; Singh, M., Gupta, A., Laxmi, A. (2017), "Striking the right chord: Signaling enigma during root gravitropism," *Frontiers in Plant Science* 8: 1304; Vandenbrink, J. P., Kiss, J. Z. (2019), "Plant responses to gravity," *Seminars in Cell & Developmental Biology* 92: 122–125.

30. Bastien, R., Bohr, T., Moulia, B., Douady, S. (2013), "Unifying model of shoot gravitropism reveals proprioception as a central feature of posture control in plants," *Proceedings of the National Academy of Sciences* 110: 755–760; Dumais, J. (2013), "Beyond the sine law of plant gravitropism," *Proceedings of the National Academy of Sciences* 110: 391–392.

31. Elhakeem, A., Markovic, D., Broberg, A., Anten, N. P. R., Ninkovic, V. (2018), "Aboveground mechanical stimuli affect belowground plant-plant communication," *PLoS ONE* 13: e0195646.

32. Falik, O., Hoffmann, I., Novoplansky, A. (2014), "Say it with flowers," *Plant Signaling & Behavior* 9: e28258.

33. Gaillochet, C., Lohmann, J. U. (2015), "The never-ending story: from pluripotency to plant developmental plasticity," *Development* 142: 2237–2249.

34. Leopold, A. C., Jaffe, M. J., Brokaw, C. J., Goebel, G. (2000), "Many modes of movement," *Science* 288: 2131–2132.

35. Trewavas, A. (2009), "What is plant behaviour?" *Plant, Cell & Environment* 32: 606–616.

36. Palacio-Lopez, K., Beckage, B., Scheiner, S., Molofsky, J.

(2015), "The ubiquity of phenotypic plasticity in plants: a synthesis," *Ecology and Evolution* 5: 3389–3400; Schlichting, C. D. (1986), "The evolution of phenotypic plasticity in plants," *Annual Review of Ecology and Systematics* 17: 667–693; Sultan, S. E. (2015), *Organism and Environment: Ecological Development, Niche Construction, and Adaptation.* Oxford: Oxford University Press.

37. Calvo, P. (2018), "Plantae." In Vonk, J., Shackelford, T. K., eds., *Encyclopedia of Animal Cognition and Behavior.* New York: Springer.

38. Segundo-Ortin, M., Calvo, P. (2021), "Consciousness and cognition in plants," *Wiley Interdisciplinary Reviews: Cognitive Science*, e1578.

39. Calvo, P., Gagliano, M., Souza, G. M., Trewavas, A. (2020), "Plants are intelligent, here's how," *Annals of Botany* 125: 11–28.

40. Baldwin, I. T. (2010), "Plant volatiles," *Current Biology* 20: R392. Current gas chromatography–mass spectrometry techniques of the sort customarily deployed by the food and beverage or perfume industries can reveal the component molecules of each different volatile organic compound; of the specific volatile concentrations that constitute the message being relayed.

41. Knudsen, J. T., Eriksson, R., Gershenzon, J., Ståhl, B. (2006), "Diversity and distribution of floral scent," *The Botanical Review* 72: 1.

42. Vivaldo, G., Masi, E., Taiti, C., Caldarelli, G., Mancuso, S. (2017), "The network of plants volatile organic compounds," *Scientific Reports* 7: 11050.

43. Another well-known case is provided by barley and thistle, where the latter does the whispering and the former the eavesdropping: Glinwood, R., Ninkovic, V., Pettersson, J., Ahmed, E. (2004), "Barley exposed to aerial allelopathy from

thistles (*Cirsium spp.*) becomes less acceptable to aphids," *Ecological Entomology* 29: 188–195.

44. Arimura, G., Ozawa, R., Shimoda, T., Nishioka, T., Boland, W., Takabayashi, J. (2000), "Herbivory-induced volatiles elicit defence genes in lima bean leaves," *Nature* 406: 512–513.

45. Passos, F. C. S., Leal, L. C. (2019), "Protein matters: ants remove herbivores more frequently from extrafloral nectary-bearing plants when habitats are protein poor," *Biological Journal of the Linnean Society* XX: 1–10.

46. Dudley, S. A., File, A. L. (2007), "Kin recognition in an annual plant," *Biology Letters* 3: 435–438; Biedrzycki, M. L., Bais, H. P. (2010), "Kin recognition: another biological function for root secretions," *Plant Signaling & Behavior* 5: 401–402; Biedrzycki, M. L., Jilany, T. A., Dudley, S. A., Bais, H. P. (2010), "Root exudates mediate kin recognition in plants," *Communicative & Integrative Biology* 3: 28–35.

47. Bais, H. P. (2015), "Shedding light on kin recognition response in plants," *New Phytologist* 205: 4–6; Crepy, M. A., Casal, J. J. (2015), "Photoreceptor-mediated kin recognition in plants," *New Phytologist* 205: 329–338.

48. Cahill Jr, J. F., McNickle, G. G., Haag, J. J., Lamb, E. G., Nyanumba, S. M., St Clair, C. C. (2010), "Plants integrate information about nutrients and neighbors," *Science* 328: 1657.

49. Delory, B. M. (2016), "Root-emitted volatile organic compounds: can they mediate belowground plant-plant interactions?" *Plant Soil* 402: 1–26; Semchenko, M., Saar, S., Lepik, A. (2014), "Plant root exudates mediate neighbour recognition and trigger complex behavioural changes," *New Phytologist* 204: 631–637; Chen, B. J. W., During, H. J., Anten, N. P. (2012), "Detect thy neighbor: identity recognition at the root level in plants," *Plant Science* 195: 157–167.

50. Dener, E., Kacelnik, A., Shemesh, H. (2016), "Pea plants show risk sensitivity," *Current Biology* 26: 1763–1767.

51. Karban, R., Orrock, J. L. (2018), "A judgement and decision-making model for plant behaviour," *Ecology* 99: 1909e1919; Gruntman, M., Groß, D., Májeková, M., Tielbörger, K. (2017), "Decision-making in plants under competition," *Nature Communications* 8: 2235; Schmid, B. (2016), "Decision-making: Are plants more rational than animals?" *Current Biology* 26: R675–R678.

52. As Roblin observes, "The first physiological experiment concerning mimosa was related in Hooke's famous *Micrographia* (1665) a century earlier: "upon touching any of the springs with leaves on, all the leaves of that sprig contracting themselves by pairs, joyned their upper superficies close together. Upon the dropping a drop of *Aqua fortis* on the sprig betwixt the leaves, all the leaves above shut presently, those below by pairs successively after and by the lower leaves on the other branches."" Roblin, G. (1979), "*Mimosa pudica*: a model for the study of the excitability in plants," *Biological Reviews* 54: 135–153.

53. Hiernaux, Q. (2019), "History and epistemology of plant behaviour: a pluralistic view?" *Synthese* 198: 3625–3650.

54. Pfeffer, W. (1873), *Physiologische untersuchungen*. Leipzig: Springer; see also Bose, J. C. (1906), *Plant Response*. London: Longmans, Green and Co.

55. Gagliano, M., Renton, M., Depczynski, M., Mancuso, S. (2014), "Experience teaches plants to learn faster and forget slower in environments where it matters," *Oecologia* 175: 63–72.

56. Tafforeau, M., Verdus, M. C., Norris, V., Ripoll, C., Thellier, M. (2006), "Memory processes in the response of plants to environmental signals," *Plant Signaling & Behavior* 1: 9–14.

57. It should be mentioned, too, that Monica's research did not actually measure growth, rather position: see Holmes, E., Gruenberg, G. (1965), "Learning in plants," *Worm Runner's Digest* 7: 9–12; Holmes, E., Yost, M. (1966), "'Behavioral' studies in the sensitive plant," *Worm Runner's Digest* 8: 38–40.

58. Gagliano, M., Vyazovskiy, V. V., Borbély, A. A., Grimonprez, M., Depczynski, M. (2016), "Learning by association in plants," *Scientific Reports* 6: 38427.

59. Darwin, *The Power of Movement in Plants*, pp. 460–461.

60. Latzel, V., Münzbergová, Z. (2018), "Anticipatory behavior of the clonal plant *Fragaria vesca*," *Frontiers in Plant Science* 9: 1847.

61. Positive results were reported in Armus, H. L. (1970), "Conditioning of the sensitive plant, Mimosa pudica," In Denny, M. R., Ratner, S. C., eds., *Comparative Psychology: Research in Animal Behavior*. Homewood, IL: Dorsey Press, pp. 597–600), and unclear results in Haney, R. E. (1969), "Classical conditioning of a plant: Mimosa pudica," *Journal of Biological Psychology* 11: 5–12; Levy, E., Allen, A., Caton, W., Holmes, E. (1970), "An attempt to condition the sensitive plant: Mimosa pudica," *Journal of Biological Psychology* 12: 86–87—for a review see Adelman, B. E. (2018), "On the conditioning of plants: A review of experimental evidence," *Perspectives on Behavior Science* 41: 431–446; Gagliano, M., Vyazovskiy, V. V., Borbély, A. A., Depczynski, M., Radford, B. (2020), "Comment on 'Lack of evidence for associative learning in pea plants'," *eLife* 9: e61141; Markel, K. (2020), "Lack of evidence for associative learning in pea plants," *eLife* 9: e57614; Markel, K. (2020), "Response to comment on 'Lack of evidence for associative learning in pea plants'," *eLife* 9: e61689 for the latest exchanges on the evidence, or lack of, for associative learning in plants.

62. Bhandawat, A., Jayaswall, K., Sharma, H., Roy, J. (2020), "Sound as a stimulus in associative learning for heat stress in Arabidopsis," *Communicative & Integrative Biology* 13: 1–5.

4: Phytonervous Systems

1. Bose, Sir J. C. (1926), *The Nervous Mechanism of Plants*. London: Longmans, Green and Co.
2. Shepherd, V. A. (2005), "From semi-conductors to the rhythms of sensitive plants: the research of J.C. Bose," *Cellular and Molecular Biology* 51: 607–19; Minorsky, P. V. (2021), "American racism and the lost legacy of Sir Jagadis Chandra Bose, the father of plant neurobiology," *Plant Signaling & Behavior* 16: 1818030.
3. Georgia O'Keeffe (1939), *Iao Valley, Maui (Papaya Tree)*, oil on canvas (Honolulu Museum of Art, gifted by the Georgia O'Keeffe Foundation). See Groake, J. L., Papanikolas, T. (eds) (2018), *Georgia O'Keeffe: Visions of Hawai'i*. New York: New York Botanical Garden.
4. Darwin, C. (1875), *Insectivorous Plants*. London: John Murray. Quoted in Volkov, A. G., ed. (2006), *Plant Electrophysiology*. Berlin: Springer.
5. Umrath, K. (1930), "Untersuchungen über Plasma und Plasmaströmung an Characeen," *Protoplasma* 9: 576–597.
6. Volkov (ed.), *Plant Electrophysiology*.
7. Li, J.-H., Fan, L. F., Zhao, D. J., Zhou, Q., Yao, J. P., Wang, Z. Y., Huang, L. (2021), "Plant electrical signals: A multidisciplinary challenge," *Journal of Plant Physiology* 261: 15341.
8. Fromm, J., Lautner, S. (2007), "Electrical signals and their physiological significance in plants." *Plant, Cell & Environment* 30: 249–257.
9. Stahlberg, R., Cleland, R. E., Van Volkenburgh, E. (2006), "Slow wave potentials—a propagating electrical signal unique to higher plants." In Baluška, F., Mancuso, S.,

Volkmann, D., eds., *Communication in Plants: Neuronal Aspects of Plant Life*. New York: Springer.

10. Baluška, F. (2010), "Recent surprising similarities between plant cells and neurons," *Plant Signaling & Behavior* 5: 87–89.

11. https://www.sciencealert.com/this-creeping-slime-is-changing-how-we-think-about-intelligence

12. Ramakrishna, A., Roshchina, V. V., eds. (2019), *Neurotransmitters in Plants: Perspectives and Applications*. Boca Raton, FL: Taylor and Francis.

13. Bouché, N., Lacombe, B., Fromm, H. (2003), "GABA signaling: a conserved and ubiquitous mechanism," *Trends in Cell Biology* 13: 607–610.

14. Bouché N., Fromm, H. (2004), "GABA in plants: just a metabolite?" *Trends in Plant Science* 9: 110–115.

15. Calvo, P. (2016), "The philosophy of plant neurobiology: A manifesto," *Synthese* 193: 1323–1343.

16. Toyota, M., Spenser, D., Sawai-Toyota, S., Jiaqi, W., Zhang, T., Koo, A. J., Howe, G. A., Gilroy, S. (2018), "Glutamate triggers long-distance, calcium-based plant defense signalling," *Science* 361: 1112–1115.

17. Brenner, E. D., Stahlberg, R., Mancuso, S., Vivanco, J., Baluška, F., Van Volkenburgh, E. (2006), "Plant neurobiology: an integrated view of plant signaling," *Trends in Plant Science* 11: 1380–1386.

18. Forde, B. G., Lea, P. J. (2007), "Glutamate in plants: metabolism, regulation, and signalling," *Journal of experimental botany* 58: 2339–2358; Baluška, F., Mancuso, S. (2009), "Plants and animals: convergent evolution in action?" In *Plant–Environment Interactions*, pp. 285–301. Berlin and Heidelberg: Springer; Baluška, F. (2010), "Recent surprising similarities between plant cells and neurons," *Plant Signaling & Behavior* 5: 87–89.

19. Morrens, J., Aydin, Ç., van Rensburg, A. J., Rabell, J. E.,

Haesler, S. (2020), "Cue-evoked dopamine promotes conditioned responding during learning," *Neuron* 106: 142–153.

20. Antoine, G. (2013), "Plant learning: an unresolved question," Master BioSciences, Département de Biologie, Ecole Normale Supérieure de Lyon.

21. Mallatt, J., Blatt, M. R., Draguhn, A., Robinson, D. G., Taiz, L. (2020), "Debunking a myth: plant consciousness," *Protoplasma* 258: 459–476.

22. Klejchova, M., Silva-Alvim, F. A., Blatt, M. R., Alvim, J. C. (2021), "Membrane voltage as a dynamic platform for spatio-temporal signalling, physiological and developmental regulation," *Plant Physiology* 185(4): 1523–1541.

23. Personal communication.

24. https://www.scientificamerican.com/article/do-plants-think-daniel-chamovitz

25. Personal communication.

26. Of course, that wasn't that. The debate continued; see Van Volkenburgh, E., Mirzaei, K., Ybarra, Y. (2021), "Understanding plant behavior: a student perspective," *Trends in Plant Science* 26: 423–425; Mallatt, J., Robinson, D. G., Draguhn, A., Blatt, M., Taiz, L. (2021), "Understanding plant behavior: a student perspective: response to Van Volkenburgh et al.," *Trends in Plant Science* 26: 1089–1090; Van Volkenburgh, E. (2021), "Broadening the scope of plant physiology: response to Mallatt et al," *Trends in Plant Science* 26: 1091–1092.

27. Machery, E. (2012), "Why I stopped worrying about the definition of life ... and why you should as well," *Synthese* 185: 145–164.

28. Miguel-Tomé, S., Llinás, R. R. (2021), "Broadening the definition of a nervous system to better understand the evolution of plants and animals," *Plant Signaling & Behavior* 10: e1927562.

29. Lucas, W. J., Groover, A., Lichtenberger, R., Furuta, K., Yadav, S.-R., Helariutta, Y., He, X.-Q., Fukuda, H., Kang, J., Brady, S. M., Patrick, J. W., Sperry, J., Yoshida, A., Lopez-Milan, A.-F., Grusak, M. A., Kachroo, P. (2013), "The plant vascular system: Evolution, development and functions," *Journal of Integrative Plant Biology* 55: 294–388.

30. Souza, G. M., Ferreira, A. S., Saraiva, G. F. R., Toledo, G. R. A. (2017), "Plant 'electrome' can be pushed towards a self-organized critical state by external cues: Evidences from a study with soybean seedlings subject to different environmental conditions," *Plant Signaling & Behavior* 12: e1290040.

31. Richard Axel, interview available on the website for the 2009 documentary *Naturally Obsessed: The Making of a Scientist*, see http://naturallyobsessed.com

32. Szent-Györgyi, A., "Electronic Mobility in Biological Processes." In Breck, A. D., Yourgrau, W., eds. (1972), *Biology, History, and Natural Philosophy*. New York: Plenum Press. Thanks to František Baluška for bringing this quote to my attention.

33. Tolman, E. C. (1958), *Behavior and Psychological Man: Essays in Motivation and Learning*. California: University of California Press. Thanks to Vicente Raja for bringing this quote to my attention.

34. Cvrčková, F., Žarský, V., Markoš, A. (2016), "Plant studies may lead us to rethink the concept of behavior," *Frontiers in Psychology* 7: 622.

35. Heras-Escribano, M., Calvo, P. (2020), "The philosophy of plant neurobiology." In Robins, S., Symons, J., Calvo, P., eds., *The Routledge Companion to Philosophy of Psychology*, pp. 529–547. London and New York: Routledge.

5: Do Plants Think?

1. Siegel, E. H., Wormwood, J. B., Quigley, K. S., Barrett, L.
    F. (2018), "Seeing what you feel: Affect drives visual per-
    ception of structurally neutral faces," *Psychological Science*
    29: 496–503.
2. The picture was first published in *Life* magazine: From James,
    R. C. (1965), Photo of a Dalmatian dog. *Life* Magazine,
    58(7), 120.
3. Gregory, R. L. (2005), The Medawar Lecture 2001:
    "Knowledge for vision: vision for knowledge," *Philosophical
    Transactions of the Royal Society B: Biological Sciences* 360:
    1231–1251.
4. Ge, X., Zhang, K., Gribizis, A., Hamodi, A. S., Martinez
    Sabino, A., Crair, M. C. (2021), "Retinal waves prime visual
    motion detection by simulating future optic flow," *Science*
    373: eabd0830.
5. See Clark, A. (1997), *Being There: Putting Brain, Body, and
    World Together Again.* Cambridge, MA: MIT Press.
6. Clark, A., Chalmers, D. (1998), "The Extended Mind,"
    Analysis 58: 7–19.
7. MacFarquhar, L. (2 Apr 2018), "The Mind-Expanding Ideas
    of Andy Clark," Annals of Thought, *New Yorker*.
8. Clark, A. (2016), *Surfing Uncertainty: Prediction, Action,
    and the Embodied Mind.* New York: Oxford University
    Press. For an introduction to predictive processing see Wiese,
    W., Metzinger, T. (2017), "Vanilla PP for philosophers: A
    primer on predictive processing." In Metzinger, T., Wiese,
    W., eds., *Philosophy and Predictive Processing* 1. Frankfurt
    am Main: MIND Group.
9. Friston, K. (2005), "A theory of cortical responses,"
    *Philosophical Transactions of the Royal Society B: Biological
    Sciences*, 360 (1456): 815–836.

10. Friston, K. (2009), "The free-energy principle: A rough guide to the brain?" *Trends in Cognitive Sciences*, 13 (7): 293–301.
11. Calvo, P., Friston, K. (2017), "Predicting green: really radical (plant) predictive processing," *Journal of the Royal Society Interface* 14: 20170096.
12. Galvan-Ampudia, C. S., Julkowska, M. M., Darwish, E., Gandullo, J., Korver, R. A., Brunoud, G. et al. (2013), "Halotropism is a response of plant roots to avoid a saline environment," *Current Biology* 23: 2044–2050; Rosquete, M. R., Kleine-Vehn, V. (2013), "Halotropism: turning down the salty date," *Current Biology* 23: R927–R929.
13. Parida, A. K., Das, A. B. (2005), "Salt tolerance and salinity effects on plants: a review," *Ecotoxicology and Environmental Safety* 60: 324–349.
14. Calvo and Friston, "Predicting green."
15. Snow, P. (2018), *Tales from Wullver's Hool: The Extraordinary Life and Prodigious Works of Jessie Saxby*. Lerwick: Shetland Times Ltd.
16. Hatfield, G. (2020), "Rationalist roots of modern psychology." In Robins, S., Symons, J., Calvo, P., eds., *The Routledge Companion to Philosophy of Psychology*. 2nd edition, London and New York: Routledge. That being said, Descartes may have been involved to some extent, however modest, in the study of plants beyond what's usually acknowledged. See Baldassarri, F. (2019), "The mechanical life of plants: Descartes on botany," *The British Journal for the History of Science* 52: 41–63.
17. Boden, M. A. (2006), *Mind as Machine: A History of Cognitive Science*, 2 vols. Oxford: Oxford University Press.
18. Fodor, J. A. (1968), *Psychological Explanation: An Introduction to the Philosophy of Psychology*. New York: Random House.
19. Marr, D. (1982), *Vision*. San Francisco: Freeman.

6: Ecological Cognition

1. Rumelhart, D. E., McClelland, J. L., PDP Research Group (1986), *Parallel Distributed Processing: Explorations in the Microstructure of Cognition*, Vol. 1. Cambridge, MA: MIT Press; Rolls, E. T., Treves, A. (1998), *Neural Networks and Brain Function*. Oxford: Oxford University Press; O'Reilly, R., Munakata, Y. (2000), *Computational Explorations in Cognitive Neuroscience*. Cambridge, MA: MIT Press; Marcus, G. F. (2001), *The Algebraic Mind: Integrating Connectionism and Cognitive Science*. Cambridge, MA: MIT Press.

2. Wilkes, M. (1975), "How Babbage's dream came true," *Nature* 257: 541–544.

3. Aiello, L. C. (2016), "The multifaceted impact of Ada Lovelace in the digital age," *Artificial Intelligence* 235: 58–62.

4. Karihaloo, B. L., Zhang, K., Wang, J. (2013), "Honeybee combs: how the circular cells transform into rounded hexagons," *Journal of the Royal Society Interface* 10: 20130299.

5. Simon, H. A. (1969), *The Sciences of the Artificial*. Cambridge, MA: MIT Press.

6. Gibson, J. J. (1979), *The Ecological Approach to Visual Perception*. Boston, MA: Houghton Mifflin.

7. Mace, W. (1977), "James J. Gibson's strategy for perceiving: Ask not what's inside your head, but what's your head inside of." In Shaw, R., Bransford, J., eds., *Perceiving, Acting, and Knowing: Towards an Ecological Psychology*. Hillsdale, NJ: Erlbaum. See also Bruineberg, J., Rietveld, E. (2019), "What's inside your head once you've figured out what your head's inside of," *Ecological Psychology*, 31:3, 198–217.

8. Chemero, A. (2011), *Radical Embodied Cognitive Science*. Cambridge, MA: MIT Press.

9. Lee, D. N., Reddish, P. L. (1981), "Plummeting gannets: A paradigm of ecological optics," *Nature* 293: 293–294.

10. Lee, D. N., Bootsma, R. J., Frost, B. J., Land, M., Regan, D. (2009). "General Tau Theory: Evolution to date," Special Issue: Landmarks in Perception, *Perception* 38: 837–858.

11. Turvey, M. T. (2018), *Lectures on Perception: An Ecological Perspective.* New York: Routledge.

12. Gibson, J. J., ed. (1947), *Motion Picture Testing and Research Report No. 7.* Washington, DC: US Government Printing Office.

13. Gibson, J. J. (1966), *The Senses Considered as Perceptual Systems.* Boston, MA: Houghton Mifflin; Gibson, *The Ecological Approach to Visual Perception.*

14. Calvo, P., Raja, V., Lee, D. N. (2017), "Guidance of circumnutation of climbing bean stems: An ecological exploration," *bioRxiv* 122358.

7: What Is It Like to Be a Plant?

1. Nagel, T. (1974), "What is it like to be a bat?" *Philosophical Review* 83: 435–450.

2. Abbott, S. (2020), "Filming with nonhumans." In Vannini, P., *The Routledge International Handbook of Ethnographic Film and Video.* Abingdon and New York: Routledge.

3. The Tree Listening Project (A. Metcalf, 2019), https://www.treelistening.co.uk, was also part of Kew Gardens' "The Secret World of Plants" exhibition from May to September 2021.

4. Jackson, F. (1982), "Epiphenomenal qualia," *Philosophical Quarterly* 32: 127–136.

5. Churchland, P. M. (1985), "Reduction, qualia, and the direct introspection of brain states," *Journal of Philosophy* 82: 8–28.

6. My thanks to Paul for reminding me of the details of this example twenty years after the fact. More in his book, Churchland, P. M. (1979), *Scientific Realism and the Plasticity of Mind*. Cambridge: Cambridge University Press (section 4, "The Expansion of Perceptual Consciousness').

7. Mather, J. A., Dickel, L. (2017), "Cephalopod complex cognition," *Current Opinion in Behavioral Sciences* 16: 131–137; Bayne, T., Brainard, D., Byrne, R. W., Chittka, L., Clayton, N., Heyes, C., Mather, J., Ölveczky, B., Shandlen, M., Suddendorf, T., Webb, B. (2019), "What is cognition?," *Current Biology* 29: R603–R622.

8. Godfrey-Smith, P. (2016), *Other Minds: The Octopus and the Evolution of Intelligent Life*. Glasgow: William Collins.

9. Dawson, J. H., Musselman, L. J., Wolswinker, P., Dorr, I. (1994), "Biology and control of *Cuscuta*," *Review of Weed Science* 6: 265–317; Gaertner, E. E. (1950), *Studies of Seed Germination, Seed Identification, and Host Relationships in Dodders, Cuscuta spp.: Memoir*. Ithaca, NY: Cornell Agricultural Experiment Station 294.

10. Runyon, J., Mescher, M., Moraes, C. D. (2006), "Volatile chemical cues guide host location and host selection by parasitic plants," *Science* 313: 1964–1967; Johnson, B. I., De Moraes, C. M., Mescher, M. C. (2016), "Manipulation of light spectral quality disrupts host location and attachment by parasitic plants in the genus *Cuscuta*," *Journal of Applied Ecology* 53: 794–803; Hegenauer, V., Slaby, P., Körner, M., Bruckmüller, J.-A., Burggraf, R., Albert, I., Kaiser, B., Löffelhardt, B., Droste-Borel, I., Sklenar, J., Menke, F. L. H., Maček, B., Ranjan, A., Sinha, N., Nürnberger, T., Felix, G., Krause, K., Stahl, M., Albert, M. (2020), "The tomato receptor CuRe1 senses a cell wall protein to identify *Cuscuta* as a pathogen," *Nature Communications* 11: 5299; Ballaré, C. L., Scopel, A. L., Roush, M. L., Radosevich, S. R. (1995), "How

plants find light in patchy canopies. A comparison between wild-type and phytochrome-B-deficient mutant plants of cucumber," *Functional Ecology* 9(6): 859–868; Benvenuti, S., Dinelli, G., Bonetti, A., Catizone, P. (2005), "Germination ecology, emergence and host detection in *Cuscuta campestris*," *Weed Research* 45: 270–278; Parise, A. G., Reissig, G. N., Basso, L. F., Senko, L. G. S., Oliveira, T. F. C., de Toledo, G. R. A., Ferreira, A. S, Souza, G. M. (2021), "Detection of different hosts from a distance alters the behaviour and bioelectrical activity of *Cuscuta racemosa*," *Frontiers in Plant Science* 12: 594195.

11. Strong, D. R. J., Ray, T. S. J. (1975), "Host tree location behavior of a tropical vine (*Monstera gigantea*) by skatotropism," *Science* 190: 804–806.

12. Price, A. J., Wilcut, J. W. (2007), "Response of ivyleaf morningglory (*Ipomoea hederacea*) to neighboring plants and objects," *Weed Technology* 21: 922–927.

13. Baillaud, L. (1962), "Mouvements autonomes des tiges, vrilles et autre organs." In Ruhland, W., ed., *Encyclopedia of Plant Physiology*, XVII: Physiology of Movements, part 2, pp. 562–635. Berlin: Springer-Verlag.

14. Vaughn, K. C., Bowling, A. J. (2011), "Biology and physiology of vines." In Janick, J., ed., *Horticultural Reviews* 38. In fact, vines are found to be associated with particular hosts: Gianoli, E. (2015), "The behavioural ecology of climbing plants," *AoB Plants* 7: plv013.

15. Parise, A. G., Reissig, G. N., Basso, L. F., Senko, L. G. S., Oliveira, T. F. C., de Toledo, G. R. A., Ferreira, A. S., Souza, G. M. (2021), "Detection of different hosts from a distance alters the behaviour and bioelectrical activity of *Cuscuta racemosa*," *Frontiers in Plant Science* 12: 594195.

16. Coined in the early 1960s by Friedrich S. Rothschild, the term "biosemiotics" has traditionally been associated with

Baltic biologist Jakob von Uexküll. Rothschild, F. S. (1962), "Laws of symbolic mediation in the dynamics of self and personality," *Annals of New York Academy of Sciences* 96: 774–784; see Kull, K., Deacon, T., Emmeche, C., Hoffmeyer, J., Stjernfelt, F. (2009), "Theses on biosemiotics: Prolegomena to a theoretical biology," *Biological Theory* 4: 167–173.

17. Jennings, H. S. (1906), *Behavior of the Lower Organisms.* New York: Columbia University Press.

18. Dexter, J. P., Prabakaran, S., Gunawardena, J. (2019), "A complex hierarchy of avoidance behaviors in a single-cell eukaryote," *Current Biology* 29: 4323–4329.

19. Uexküll, J. (1921), *Umwelt und Innenwelt der Tiere.* 2nd edition. Berlin: Springer; Uexküll, J. (1940, 1982), "The theory of meaning," *Semiotica* 42: 25–82.

20. Krampen, M. (1981), "Phytosemiotics," *Semiotica* 36: 187–209.

21. Montgomery, S. (1991), *Walking with the Great Apes: Jane Goodall, Dian Fossey, Biruté Galdikas.* Boston, MA: Houghton Mifflin.

22. Gibson, J. J. (1979), *The Ecological Approach to Visual Perception.* Boston, MA: Houghton Mifflin.

23. Gibson, J. J. (1966), *The Senses Considered as Perceptual Systems.* Boston, MA: Houghton Mifflin.

24. Raja, V. (2018), "A theory of resonance: Towards an ecological cognitive architecture," *Minds & Machines* 28: 29–51.

25. Michaels, C., Carello, C. (1981), *Direct Perception.* Englewood Cliffs, NJ: Prentice-Hall.

26. Although see Baluška, F., Mancuso, S. (2016), "Vision in plants via plant-specific ocelli?" *Trends in Plant Science* 21: 727–730. For an artistic rendering of the working hypothesis of plant vision, see Sarah Abbott's short film *Gestures toward Plant Vision*: https://www.youtube.com/watch?v=D5HTR2QfTkc

27. Vandenbrink, J. P., Kiss, J. Z. (2019), "Plant responses to gravity," *Seminars in Cell & Developmental Biology* 92: 122–125.

28. Aliperti, J. R., Davis, B. E., Fangue, N. A., Todgham, A. E., Van Vuren, D. H. (2021), "Bridging animal personality with space use and resource use in a free-ranging population of an asocial ground squirrel," *Animal Behaviour* 180: 291–306.

29. Barrett, L. P., Benson-Amram, S. (2021), "Multiple assessments of personality and problem-solving performance in captive Asian elephants (*Elephas maximus*) and African savanna elephants (*Loxodonta africana*)," *Journal of Comparative Psychology* 135: 406–419.

30. Reed-Guy, S., Gehris, C., Shi, M., Blumstein, D. T. (2017), "Sensitive plant (*Mimosa pudica*) hiding time depends on individual and state," *PeerJ Life and Environment* 5: e3598.

31. Kaminski, J., Waller, B. M., Diogo, R., Hartstone-Rose, A., Burrows, A. M. (2019), "Evolution of facial muscle anatomy in dogs," *Proceedings of the National Academy of Sciences* 116: 14677–14681.

32. Hasing, T., Rinaldi, E., Manrique, S., Colombo, L., Haak, D. C., Zaitlin, D., Bombarely, A. (2019), "Extensive phenotypic diversity in the cultivated Florist's Gloxinia, *Sinningia speciosa* (Lodd.) Hiern, is derived from the domestication of a single founder population," *Plants, People, Planet* 1: 363–374.

33. Wu, D., Lao, S., Fan, L. (2021), "De-domestication: An extension of crop evolution," *Trends in Plant Science* 26: 560–574; Scossa, F., Fernie, A. R. (2021), "When a crop goes back to the wild: Feralization," *Trends in Plant Science* 26: 543–545.

34. Spengler, R. N. (2020), "Anthropogenic seed dispersal: Rethinking the origins of plant domestication," *Trends in Plant Science* 25: 340–348; Spengler, R. N., Petraglia, M., Roberts, P., Ashastina, K., Kistler, L., Mueller, N. G., Boivin, N. (2021), "Exaptation traits for megafaunal

mutualisms as a factor in plant domestication," *Frontiers in Plant Science* 12: 43.

8: Plant Liberation

1. Mallatt, J., Blatt, M. R., Draguhn, A., Robinson, D. G., Taiz, L. (2020), "Debunking a myth: Plant consciousness," *Protoplasma* 258: 459–476.
2. Segundo-Ortin, M., Calvo, P. (2021), "Consciousness and cognition in plants," *Wiley Interdisciplinary Reviews: Cognitive Science* e1578.
3. Anderson, D. J., Adolphs, R. (2014), "A framework for studying emotions across species," *Cell* 157: 187–200.
4. Barr, S., Laming, P. R., Dick, J. T. A., Elwood, R. W. (2008), "Nociception or pain in a decapod crustacean?" *Animal Behaviour* 75: 745–751.
5. Singer, P. (1975; 2009), *Animal Liberation: The Definitive Classic of the Animal Movement*. Updated edition, New York: HarperCollins.
6. Key, B. (2015), "Fish do not feel pain and its implications for understanding phenomenal consciousness," *Biology and Philosophy* 30: 149–165; Key, B. (2016), "Why fish do not feel pain," *Animal Sentience* 3: 1.
7. Dawkins, R. (1986), *The Blind Watchmaker*, p. 37. New York: Norton. From Lent, J. (2021), *The Web of Meaning*. London: Profile Books.
8. Woodruff, M. (2018), "Sentience in fishes: more on the evidence," *Animal Sentience* 2: 16. Sentience in fishes would be equivalent to what Feinberg and Mallatt call sensory consciousness (Feinberg, T. E., Mallatt, J. M. (2016), *The Ancient Origins of Consciousness: How the Brain Created Experience*. Cambridge, MA: MIT Press), Merker calls core consciousness (Merker, B. (2007), "Consciousness without a cerebral cortex: A challenge for neuroscience and medicine,"

Behavioral and Brain Sciences 30: 63–81), and Edelman calls primary consciousness (Edelman, G. M. (2003), "Naturalizing consciousness: a theoretical framework," *Proceedings of the National Academy of Sciences* 100: 5520–5524).

9. Portavella, M., Torres, B., Salas, C. (2004), "Avoidance response in goldfish: emotional and temporal involvement of medial and lateral telencephalic pallium," *Journal of Neuroscience* 24: 2335–2342. Vargas, J. P., López, J. C., Portavella, M. (2009), "What are the functions of fish brain pallium?" *Brain Research Bulletin* 79: 436–440.

10. Calvo, P., Sahi, V. P., Trewavas, A. (2017), "Are plants sentient?" *Plant, Cell & Environment* 40: 2858–2869; Baluška, F. (2010), "Recent surprising similarities between plant cells and neurons," *Plant Signaling & Behavior* 5: 87–89; Baluška, F., Mancuso, S. (2013), "Ion channels in plants. From bioelectricity to behavioural actions," *Plant Signaling & Behaviour* 8: e23009; Arnao, M. B., Hernández-Ruiz, J. (2015), "Functions of melatonin in plants: a review," *Journal of Pineal Research* 59: 133–150.

11. Lew, T. T. S, Koman, V. B., Silmore, K. S., Seo, J. S., Gordiichuk, P., Kwak, S.-Y., Park, M., Ang, M. C., Khong, D. T., Lee, M. A., Chan-Park, M. B., Chua, N.-M., Strano, M. S. (2020), "Real-time detection of wound-induced H_2O_2 signalling waves in plants with optical nanosensors," *Nature Plants* 6: 404; Zhang, L., Takahashi, Y., Hsu, P.-K., Hannes, K., Merilo, E., Krysan, P. J., Schroeder, J. I. (2020), "FRET kinase sensor development reveals SnRK2/OST1 activation by ABA but not by MeJA and high CO_2 during stomatal closure," *eLife* 9: e56351.

12. Taylor, J. E. (1891), *The Sagacity and Morality of Plants. A Sketch of the Life and Conduct of the Vegetable Kingdom.* New edition, London: Chatto & Windus. My thanks to Alan Costall for these references.

13. Emeritus, Brooklyn College and Graduate Center of the City University of New York.
14. Reber, A. S. (2019), *The First Minds: Caterpillars, 'Karyotes and Consciousness.* New York: Oxford University Press.
15. Baluška, F., Reber, A. (2019), "Sentience and consciousness in single cells: How the first minds emerged in unicellular species," *BioEssays* 41: 1800229.
16. Mitchell, A., Romano, G. A., Groisman, B., Yona, A., Dekel, E., Kupiec, M., Dahan, O., Pilpel, Y. (2009), "Adaptive prediction of environmental changes by microorganisms," *Nature* 460: 220–224; Tagkopoulos, I., Liu, Y.-C., Tavazoie, S. (2008), "Predictive behavior within microbial genetic networks," *Science* 320: 1313–1317; Calvo, P., Baluška, F., Trewavas, A. (2021), "Integrated information as a possible basis for plant consciousness," *Biochemical and Biophysical Research Communications* 564: 158–165.
17. Reber, A. S. (2016), "Caterpillars, consciousness and the origins of mind," *Animal Sentience* 1: 11(1). Reber is not the first to have a broad view of consciousness in which movement is a prerequisite; see Klein and Barron on insect experience: "The life of a tree does not demand high speed general purpose perception, flexible planning and precisely controlled action" (Klein, C., Barron, A. B. (2016), "Insects have the capacity for subjective experience," *Animal Sentience* 9).
18. Calvo, P. (2018), "Caterpillar/basil-plant tandems," *Animal Sentience*, 11 (16).
19. Reber, A. S. (2018), "Sentient plants? Nervous minds?" *Animal Sentience* 11(17).
20. Reber (2019), *The First Minds.*
21. Stern, P. (2021), "The many benefits of healthy sleep," *Science* 374: 6567.
22. Taton, A., Erikson, C., Yang, Y., Rubin, B. E., Rifkin, S. A., Golden, J. W., Golden, S. S. (2020), "The circadian clock

and darkness control natural competence in cyanobacteria,"
Nature Communications 11: 1–11.

23. Nath, R. D., Bedbrook, C. N., Abrams, M. J., Basinger,
T., Bois, J. S., Prober, D. A., Sternberg, P. W., Gradinaru,
V., Goentoro, L. (2017), "The jellyfish *Cassiopea* exhibits a
sleep-like state," *Current Biology* 27: 2984–2990.

24. Leung, L. C. (2019), "Neural signatures of sleep in zebra-
fish," *Nature* 571: 198–204; Kupprat, F., Hölker, F., Kloas,
W. (2020), "Can skyglow reduce nocturnal melatonin
concentrations in Eurasian perch?" *Environmental Pollution*
262: 114324.

25. Beverly, D. P. (2019), "Hydraulic and photosynthetic
responses of big sagebrush to the 2017 total solar eclipse,"
Scientific Report 9: 8839.

26. Baluška. F., Yokawa, K. (2021), "Anaesthetics and plants:
from sensory systems to cognition-based adaptive behaviour," *Protoplasma* 258: 449–454.

27. Tononi, G. (2004), "An information integration theory of
consciousness," *BMC Neuroscience* 5: 1–22; Tononi, G.
(2008), "Consciousness as integrated information: A provi-
sional manifesto," *Biological Bulletin* 215: 216–242; Tononi,
G., Koch, C. (2015), "Consciousness: Here, there and everywhere?" *Philosophical Transactions of the Royal Society B:
Biological Sciences* 370: 20140167.

28. Tononi, G., Boly, M., Massimini, M., Koch, C. (2016),
"Integrated information theory: From consciousness to
its physical substrate," *Nature Reviews Neuroscience*
17: 450–461.

29. Mediano, P., Trewavas, A., Calvo, P. (2021), "Information
and integration in plants. Towards a quantitative search
for plant sentience," *Journal of Consciousness Studies*
28: 80–105; Calvo, P., Baluška, F., Trewavas, A. (2021),
"Integrated information as a possible basis for plant

consciousness," *Biochemical and Biophysical Research Communications* 564: 158–165.

30. Borisjuk, L., Rolletschek, H. Neuberger, T. (2012), "Surveying the plant's world by magnetic resonance imaging," *The Plant Journal* 70: 129–146; Hubeau, M., Steppe, K. (2015), "Plant-PET scans: In vivo mapping of xylem and phloem functioning," *Trends in Plant Sciences* 20: 676–685; Jahnke, S. (2009), "Combined MRI–PET dissects dynamic changes in plant structures and functions," *The Plant Journal* 59: 634–644.

31. Massimini, M., Boly, M., Casali, A., Rosanova, M., Tononi, G. (2009), "A perturbational approach for evaluating the brain's capacity for consciousness." In Laureys, S. et al., eds., *Progress in Brain Research* 177, pp. 201–214. Amsterdam: Elsevier; Massimini, M., Tononi, G. (2018), *Sizing Up Consciousness: Towards an Objective Measure of the Capacity for Experience.* Oxford: Oxford Scholarship Online.

32. Ludwig, D., Hilborn, R., Walters, C. (1993), "Uncertainty, resource exploitation, and conservation: lessons from history," *Science* 260: 17–36.

9: Green Robots

1. Frazier, P. A., Jamone, L., Althoefer, K., Calvo, P. (2020), "Plant bioinspired ecological robotics," *Frontiers in Robotics and AI* 7: 79; Lee, J., Calvo, P. (2022), "Enacting plant-inspired robotics," *Frontiers in Neurorobotics* 15: 772012.

2. Ozkan-Aydin, Y., Murray-Cooper, M., Aydin, E., McCaskey, E. N., Naclerio, N., Hawkes, E. W., Goldman, D. I. (2019), "Nutation aids heterogeneous substrate exploration in a robophysical root." In *2nd IEEE International Conference on Soft Robotics (RoboSoft),* pp. 172–177.

3. Mazzolai, B., Walker, I., Speck, T. (2021), "Generation

GrowBots: Materials, mechanisms, and biomimetic design for growing robots," *Frontiers in Robotics and AI* 8; Taya, M., Van Volkenburgh, E., Mizunami, M., Nomura, S. (2016), *Bioinspired Actuators and Sensors*. Cambridge University Press. This section was inspired by Barbara Mazzolai's GrowBot project at the Italian Institute of Technology in Genoa, which "proposes a disruptively new paradigm of movement in robotics inspired by the moving-by-growing abilities of climbing plants:" https:// growbot.eu

4. Yoo, C. Y., He, J., Sang, Q., Qiu, Y., Long, L., Kim, R. J., Chong, E. G., Hahm, J., Morffy, N., Zhou, P., Strader, L. C., Nagatani, A., Mo, B., Chen, X., Chen, M. (2021), "Direct photoresponsive inhibition of a p53-like transcription activation domain in PIF3 by *Arabidopsis* phytochrome B," *Nature Communications* 12: 1–16; Willige, B. C., Zander, M., Yoo, C. Y., Phan, A., Garza, R. M., Trigg, S. A., He, Y., Nery, J. R., Chen, H., Chen, M., Ecker, J. R., Chory, J. (2021), "Phytochrome-interacting factors trigger environmentally responsive chromatin dynamics in plants," *Nature Genetics* 53: 955–961.

5. https://forum.frontiersin.org/speakers

6. Terrer, C. et al. (2019), "Nitrogen and phosphorus constrain the CO_2 fertilization of global plant biomass," *Nature Climate Change* 9: 684–689; Obringer, R., Rachunok, B., Maia-Silva, D., Arbabzadeh, M., Nateghi, R., Madani, K. (2021), "The overlooked environmental footprint of increasing Internet use," *Resources, Conservation and Recycling* 167: 105389.

7. Overpeck, J. T., Breshears, D. D. (2021), "The growing challenge of vegetation change," *Science* 372: 786-787; Alderton, G. (2020), "Challenges in tree-planting programs," *Science* 368: 616.8.

8. Produced by Michael Moore and directed by Jeff Gibbs, the eco-documentary was removed from YouTube. The film is still available at https://planetofthehumans.com

9. Lawrence, N., Calvo, P. (2022), "Learning to see 'green' in an ecological crisis." In Weir, L. ed., *Philosophy as Practice in the Ecological Emergency: An Exploration of Urgent Matters.* Berlin: Springer (in press); Segundo-Ortin, M., Calvo, P. (2019), "Are plants cognitive? A reply to Adams," *Studies in History and Philosophy of Science* 73: 64–71; Segundo-Ortin, M., Calvo, P. (2021), "Consciousness and cognition in plants," *Wiley Interdisciplinary Reviews: Cognitive Science*, e1578; Baluška, F., Mancuso, S. (2020), "Plants, climate and humans: plant intelligence changes everything," *EMBO Reports* 21(3): e50109; Calvo, P., Baluška, F., Trewavas, A. (2021), "Integrated information as a possible basis for plant consciousness," *Biochemical and Biophysical Research Communications* 564: 158–165; Trewavas, A., Baluška, F., Mancuso, S., Calvo, P. (2020), "Consciousness facilitates plant behavior," *Trends in Plant Science* 25: 216–217.

10. The title of this section is derived from GrowBot: Towards a new generation of plant-inspired artefacts (https://growbot. eu), a European project funded under the EU Future and Emerging Technologies (FET) programme.

11. Song, Y., Dai, Z., Wang, Z., Full, R. J. (2020), "Role of multiple, adjustable toes in distributed control shown by sideways wall-running in geckos," *Proceedings of the Royal Society B: Biological Sciences* 287: 20200123.

12. https://news.mit.edu/2019/mit-mini-cheetah-first-four-legged-robot-to-backflip-0304

13. Hawkes, E. W., Majidi, C., Tolley, M. T. (2021), "Hard questions for soft robotics," *Science Robotics* 6: eabg6049.

14. Isnard, S., Silk, W. K. (2009), "Moving with climbing plants from Charles Darwin's time into the 21st century," *American*

Journal of Botany 96: 1205–1221; Gerbode, S. J. et al.
(2012), "How the cucumber tendril coils and overwinds,"
Science 337: 1087.

15. Yang, M., Cooper, L. P., Liu, N., Wang, X., Fok, M. P.
(2020), "Twining plant inspired pneumatic soft robotic spiral
gripper with a fiber optic twisting sensor," *Optics Express*
28: 35158–35167.

16. Frazier, P. A., Jamone, L., Althoefer, K., Calvo, P. (2020),
"Plant bioinspired ecological robotics," *Frontiers in Robotics
and AI* 7: 79.

17. Hawkes, E. W., Blumenschein, L. H., Greer, J. D., Okamura,
A. M. (2017), "A soft robot that navigates its environment
through growth," *Science Robotics* 2: eaan3028. See also Del
Dottore, E., Mondini, A., Sadeghi, A., Mazzolai, B. (2019),
"Characterization of the growing from the tip as robot loco-
motion strategy. *Frontiers in Robotics and AI* 6: 45; Laschi,
C., Mazzolai, B. (2016), "Lessons from animals and plants:
The symbiosis of morphological computation and soft robot-
ics," *IEEE Robotics & Automation Magazine* 23: 107–114;
Sadeghi, A., Mondini, A., Del Dottore, E., Mattoli, V.,
Beccai, L., Taccola, S., Lucarotti, C., Totaro, M., Mazzolai,
B. (2016), "A plant-inspired robot with soft differential bend-
ing capabilities," *Bioinspiration & Biomimetics* 12: 015001.

18. http://www.danielspoerri.org/englisch/home.htm

19. http://www.rondeeleieren.nl

20. Terrill, E. C. (2021), "Plants, partial moral status, and practical ethics," *Journal of Consciousness Studies* 28: 184–209.

21. Weir, L. (2020), *Love is Green: Compassion in Response to
the Ecological Emergency*. Wilmington, DE: Vernon Press.

22. Darwin, C. (1871/1981), *The Descent of Man, and
Selection in Relation to Sex*. Princeton, NJ: Princeton
University Press.

23. Sorabji, R. (1995), "Plants and animals." In *Animal Minds*

and Human Morals: The Origins of the Western Debate. Ithaca, NY: Cornell University Press.

24. Henkhaus, N. et al. (2020), "Plant science decadal vision 2020–2030: Reimagining the potential of plants for a healthy and sustainable future," *Plant Direct* 4: e00252.
25. Koechlin, F. (2008), "The dignity of plants," *Plant Signaling & Behavior* 4: 78–79.

Epilogue: The Hippocampus-Fattening Farm

1. Moore, J. R. (1982), "Charles Darwin lies in Westminster Abbey," *Biological Journal of the Linnean Society* 17: 97–113.
2. https://youtu.be/iG9CE55wbtY

PICTURE CREDITS

INDEX

photography
  pinhole cameras 52–3, 53
  time-lapse photography
    sequences 16, 29, 53–5,
    56–7, 62–3, 65–6, 167,
    168, 178
photosynthesis 12, 35, 39, 72,
  75, 78, 87, 98, 176, 193, 195,
  203, 209
Physarum polycephalum 40
physiological adjustments 72–3,
  82, 120, 121, 122–4
phytoethics 181, 183, 199–202,
  216–20
phytomelatonin 17
phytonervous systems 103, 107,
  118, 124–5, 161
phytopersonalities 175–8
phytosemiotics 169
phytosentience 128, 183, 196, 198,
  199–200, 215, 218, 219, 220
pigeons 147
pineal gland 17
pinhole cameras 52–3, 53
The Planet of the Humans
  (documentary) 210
planetary rovers 203–6
plant blindness 26–9, 31, 35, 208
plant neurobiology 1, 94, 96–8,
  99, 102, 104–9
  academic turf wars 102–4
  broadening the meaning of
    "neurobiology" 107–8
  interdisciplinary mix of ideas
    110–11
  phytonervous systems 103, 107,
    118, 124–5, 161
  semantic concerns 104, 105,
    106, 107
  working hypothesis of 110

  see also electrophysiology
Plant Physiology (Taiz and
  Zeiger) 97
plant psychology 103, 125–7
Plant Science Research Network
  219
plasmodium 40
Plato 36, 219
pleasure 185
poison ivy 29
Pollan, Michael 61–2, 89
pollen 36, 37, 38, 69
  production 77
pollinators 37–8, 39, 55, 74, 77,
  96, 178, 179
pollinium 55–6
pool (game) 137, 138
Positron Emission Tomography
  (PET) 198
potential hosts, detecting 162–5,
  166
*The Power of Movement in
  Plants* (Darwin) 70–1,
  147–8, 193
"Praying Palm of Faridpur" 59
preconception 115–16
predatory insects, attracting 84
predictive processing 118–20
prokaryotes 193, 203
proprioception 79
protein synthesis 122
Protista 98, 168
protozoa 190
psychology 127, 128
  plant psychology 103, 125–7
  psychology–physiology
    relationship 123, 128–9,
    130
pumpkins 97
Pythagoras 219

ABOUT THE AUTHORS

Paco Calvo is Professor of Philosophy of Science at the Minimal Intelligence Lab (MINT Lab) in the University of Murcia, Spain, where his research is primarily in exploring and experimenting with the possibility of plant intelligence. In his research at MINT Lab, he studies the ecological basis of plant intelligence by conducting experimental studies at the intersection of plant neurobiology and ecological psychology. He has given many talks on the topic of plant intelligence to academic and non-academic audiences around the world in the last decade.

Natalie Lawrence is a writer and illustrator with a PhD and MSc in the History and Philosophy of Science and a MCantab in Zoology from the University of Cambridge. Her work has appeared in *BBC Wildlife*, *Aeon Magazine* and *Public Domain Review*. She has been a TEDx speaker and appeared on BBC *Woman's Hour*.